Quantum Immortality
The Hypersoul And Afterlife

By

Rob Shelsky

Quantum Immortality
The Hypersoul
And
Afterlife

PUBLISHED BY:
GKRS PUBLICATIONS
Copyright © 2021 by Rob Shelsky

All rights reserved. Without limiting the rights under copyright reserved above, no part of this publication may be reproduced, stored in or introduced into a retrieval system, or transmitted, in any form, or by any means (electronic, mechanical, photocopying, recording, or otherwise) without the prior written permission of both the copyright owners and the above publisher of this book.

This is a work of nonfiction. The authors acknowledge the trademarked status and trademark owners of various products referenced in this work of nonfiction, which have been used without permission. The publication/use of these trademarks is not authorized, associated with, or sponsored by the trademark owners. All quotations and/or related materials/images are referenced either in the body of this book itself, or referenced at the end. All images are in the public domain.

Edition License Notes

This eBook is licensed through GKRS Publications for your personal enjoyment only. This eBook may not be re-sold or given away to other people. If you would like to share this book with another person, please purchase an additional copy for each person you share it with. If you're reading this book and did not purchase it, or it was not purchased for your use only, then you should purchase your own copy. Thank you for respecting the author's work.

* * * * *

DEDICATIONS

There is one person I would especially like to thank, because she has been such a good friend and loyal supporter for so many years:

Diane Elizabeth Powell

I wish to acknowledge you for your loyalty, dedication, mountains of help, and always just being there for me. Again, thank you, so very much, my darling woman!

~ And Always For ~

George Kempland
In Memoriam
1929—2013

I will never forget you. May we meet again, somewhere, sometime…some when.

* * * * *

Contents

Introduction 1
PART 1—THE HUMAN MIND 3
Chapter 1—"The Country of the Mind"-- Origin of Consciousness 4
Chapter 2—Consciousness Overall 16
Chapter 3—The Power Of Belief 25
PART 2—ODD TYPES OF IMMORTALITY 36
CHAPTER 4—Parallel Worlds or Realities 37
CHAPTER 5—Inflation Theory 39
CHAPTER 6—An Infinite Universe 41
CHAPTER 7—String Theory 45
CHAPTER 8—Mbrane Theory 47
CHAPTER 9—Black Hole Theory of Universe Creation 49
CHAPTER 10—Simulation Theory 52
PART 3--THE QUANTUM SOUL 58
CHAPTER 11—A Quantum Consciousness Instead 59
CHAPTER 12—Quantum Ghosts and Quantum Souls? 67
CHAPTER 13—Quantum Suicide or Immortality 72
CHAPTER 14—Are All Probabilities Carried Out? 76
CHAPTER 15—Evidence in The Body for Life After Death? 79
PART 4—THE HYPERSOUL 81
CHAPTER 16—The Hypersoul 82
CHAPTER 17—A Conscious Universe? 85
CHAPTER 18—Evidence for Near Death Experiences? 90
CHAPTER 19—Evidence for Reincarnation? 93
CONCLUSION 99
REFERENCES 101

Introduction

Death, is it the end of everything? Once we die, is that it? Are we then condemned forever to oblivion? Do we just cease to be altogether, eventually to be forgotten by the rest of humanity, becoming just so much dust upon the ground for others to tread upon? That's the question that has haunted humanity since we became conscious beings ages ago.

Down through the seemingly endless millennia since then, we have wondered about what might come after life, if anything at all. The mere thought of death scares us, frightens most of us. One of, if not the most important cries one will ever hear from another is: "I don't want to die!"

The truth is we don't. We don't want to die, at least, not most of us. So we cling to whatever might promise us that there is something more, and so for countless generations, we have relied upon various religions, all of which, seem to promise an afterlife of one sort or another. Whether it is the fiery hell of brimstone or a cloud-strewn heaven, as with the Christian churches, or paradise, as with the Muslim religion, or simply the eventual and ultimate state of Nirvana, as with Hinduism or Buddhism, they all tell us that there is something more after we die, that we go on in some way or another.

As comforting as this approach may be, many now no longer choose to believe in these religions. We have a growing number of people every year who profess to be atheists or at least, agnostics and this is taking place in every developed country on the planet.

In America, for example, Christianity's numbers are declining and quickly. In just ten years, they have declined over 11 percent. Christianity is shrinking, and not just here, but in almost every developed country. The same is true for many other major religions.

As countries move from being poor, third-world countries to developed countries, the percentage of those who profess strong religious beliefs always goes down. The United States, oddly, was one of the last to see this phenomenon develop, but we are catching up with Europe and elsewhere quickly! It is estimated that by the end of this century, only a minority of people will claim adherence to religions in general, and Christianity in particular. Again, it isn't just Christianity that faces this now-constant erosion of its base population, but all the other religions, as well, except for a tiny few, and they are small in numbers of members to begin with and more often than not, as with Scientology are given only cult status, and not even that in some countries.

So what's happening? Why are people becoming atheists and agnostics, where once they were church-going "believers?" Well, as education progresses, as people have more comfortable and high-tech lives, they just seem to let go of their belief systems with regard to religion. There are many reasons o-called experts give for this phenomenon, but one thing seems irrefutable. As standards of living increase in a country, the proportion of church goers goes down.

In England, for example, only 7 percent of the population still attend Church of England church services on a regular basis. That number is higher for Muslims and Hindus, those who still believe, but as second, third, and fourth generations of Muslim and Hindu young people mature, their attachment to their religion goes down, as well, with many paying little more than lip service to their faiths, and many others not even bothering to do that much.

So what is the current situation? A growing number of people who don't believe in any god, and a shrinking number who do, seems to be the ongoing process we are seeing as a worldwide civilization. And with atheism comes a total disbelief in the afterlife, as well as any sort of god.

Or does it? Science now seems to be rising to the challenge of this disbelief, as more and more researchers are beginning to question just what consciousness is, the nature of it, what causes it, and what it might do or become upon the death of the individual host of such a consciousness. Scientists wonder if consciousness dies with the body, or somehow goes on, and whether there is more to the universe in that regard than we thought. Some scientists even wonder if consciousness creates the universe, while still others think the universe may be conscious!

So is the afterlife just an old belief system headed for the rubbish bin of history? Alternatively, is there something that happens after life to the consciousness of each of us? In the following chapters we will delve into this idea of a life after death. We will discuss quantum immortality. Does it exist? Is it real? We will also discuss new evidence that points in some other directions, as well, with regard to a number of other topics with dealing with an afterlife. Is there evidence for such a thing at all, for the continuation of our consciousness after our bodies have died? Well, let's find out.

One last thing; try to keep an open mind with regard to some of these topics. At least consider them, because as Shakespeare's Hamlet mentions:

> "There are more things in heaven and earth, Horatio,
> Than are dreamt of in your philosophy.'

PART 1—THE HUMAN MIND

Chapter 1—"The Country of the Mind"-- Origin of Consciousness

*"O, what a world of unseen visions and heard silences,
this insubstantial country of the mind!
What ineffable essences,
these touchless rememberings and unshowable reveries!"*

— Julian Jaynes,
*The Origin of Consciousness in the
Breakdown of the Bicameral Mind*

So where does consciousness come from? Just what is the "country of the mind," is it an illusion, a mere byproduct of the brain, or is it real, and if so, how does it exist? What is it, exactly? That's just a few of the questions researchers are trying to answer, yet nobody seems to know for sure at this point, just what those answers are. Before we get to trying to reach such answers ourselves, first we should try to decide what consciousness is. In this regard, there are three main theories for consciousness:

Idealist View of Consciousness. This is the idea that the human consciousness is an entity in its own right, compete unto itself and ultimately not dependent on the human body. Perhaps with regard to this approach, one should then better define human consciousness as the "soul." As René Descartes stated, even if everything is an illusion around us and existence is not real at all, but just the phony creation of some "demon," (or whatever), our consciousness is exempt from this. It is still *not* the creation of that demon or its plaything, and the human consciousness therefore, is not just a result of being another part of a totally false existence, even if the world around it is.

Even in the movie, *The Matrix*, though all reality was a sham, an illusion, the minds of the people involved were still real enough, although they had been tricked into thinking that all around them was real when it was not. Although there were demons (the machines) involved in faking of humans' reality, there was still one thing that was real, the people and their consciousness's were.

So in at least one respect, although the demon (or whatever agency—programmer of a matrix, perhaps?) can create a false reality, gives us a sense of realness and existence falsely through the use of tricking us with our five senses, there is one thing s/he cannot do. S/he cannot make us consider the idea of that existence. That is reserved for us alone, our consciousness or soul according to Descartes.

So, even though all reality, all existence may be illusion, how does that account for one's inner ability to consider their own existence and place in the universe whether that universe is real or fake? Remember, Descartes and his, *"I think, therefore I am."*

We can think. We can consider things, ponder issues. We can contemplate and then persevere for answers to various philosophical questions. We can even question if the "demon" exists, even as we can question everything about existence in general. So what does this mean? How do we resolve this idea of consciousness versus reality? Is it part of it? Is it separate from it? Is it just an illusion?

Well, Descartes answered this by saying that consciousness does exist and therefore, must be divorced from the physical existence, must be totally separate from the physical body in order to be able to achieve such capability as to contemplate the nature of all else. Descartes feels it is the only way that consciousness could then consider its own existence, its own or others for that matter. Remember, not only can you ponder your own existence, but you can even ponder the fact that you are pondering it!

Therefore, the physical part of us may be part of that "existence," (real or contrived, as in *The Matrix*) but the consciousness is not. Thus, again, Descartes states that consciousness and the human body are two separate things, that there is a "dual nature" to humans, their physical selves and conscious selves.

Once more, whether you call it a soul, spirit, or whatever you wish, it is Descartes belief that based on logic, such a thing must exist in its own right. He points out that humans have or seem to have free will. Moreover, as we shall later see, the idea of a "dual" nature to humanity, goes very well with the idea of the dual nature of particles, as set forth in the basics of quantum physics.

Another noted and contemporary philosopher, Deepak Chopra, argues much the same thing in a different way. He says that we are creatures of thought, in that we constantly, from moment to moment, have thoughts.

However, and this is a critical point in his approach to the issue, Mr. Chopra also argues that (and this is the really important part) we are not our thoughts. They are not the sum of our consciousness. Instead, our true selves are the thinker of those thoughts. This means that he makes our consciousness, our innate self, again, not just the sum total of our thoughts, but rather something or someone else, a voice as it were who is the actual hidden thinker inside us that is the source of all those thoughts. And that "thinker" has free will. It can choose. It can choose to have thoughts on different subjects, for instance. It can fantasize. It can daydream. And it exists. The deep, inner self is the creator of all those thoughts we have. It is the true us.

As an example, what you see through your eyes is only a recreation in your mind of what is truly "out there," the physical reality all around us. The real landscape is not the recreation of it that you have in your mind, but is a real thing "out there," outside of us and so separate from us.

We simply use our eyes to transmit light bouncing off that scene to capture it and then send a signal up our optic nerve to our brain, which then recreates the scene for us inside our brain. But the interior scene is not the same as the exterior scene. Rather, it is a mental recreation of it.

Don't believe this? Then ask yourself how a colorblind person can perceive what they see differently from how you might perceive it as? Or for those who have lost the ability to smell who can't enjoy the odor of a rose and never have. Their reality is perceptibly altered from our own. Their mental recreation of "out there," differs in some respect from ours because of their physical capabilities being different from ours.

This idea can go further. There are people who are born with (or somehow acquire) an ability to actually perceive the world on a basic level, but differently than most others do. Some people "can see" words as they are spoken. Others "see" sounds as shapes and colors floating in the air. This is referred to as "synesthesia." It can come in many forms, with some people able to "taste" or smell" words, sounds, etc.

The current theory offered to explain this is that when we are very young, our brains are not yet "hardwired" completely to our senses. Thus, some children can smell or taste pictures in books, or taste or smell written words, while others can do the same with sounds, etc. It is almost as if their brain is temporarily cross-wired. This ability/condition usually goes away on its own by about the age of six or seven, but for a very few, it carries on into adulthood as a permanent thing.

I, myself, had synesthesia as a child up until about the age of seven. I could smell and taste pictures in books, as well as perceive certain odors (not real ones in the ordinary sense one supposes) upon entering a place. On a trip to grandfather's apartment in Massachusetts when I was very young, I remarked to my father upon entering that Grandpa's place smelled like cinnamon oatmeal. My grandfather had not had oatmeal in weeks, and my father was actually annoyed at the comment. I quickly learned never to voice such things in an adult's presence.

Of course, by about the age of seven, the ability faded away, and just exactly when I'm not certain, but one day, I realized I couldn't do it anymore. Still, I think it did make me develop a love for words, written and spoken, and to this day I feel I am a writer because of that.

However, my point here is that people, including myself when I was very young, really did (and some still do even when they are older) perceive reality in a fundamentally different way from others. So perception and consciousness aren't the same thing, although they affect each other. How one perceives is often the basis for how one thinks. But perceptions of what is "real" what is "reality" can and do vary.

Therefore, thoughts are not an end in themselves, are not our consciousness, but merely the mental expression of something deeper. Thoughts then, are just the manifestation of a deeper consciousness according to Chopra, and that being is the "thinker of those thoughts." So in his estimation, our thoughts are not us, do not create, form, or act as the sum total of us, but rather there is a deeper self and that is the true "thinker," the consciousness (or soul, as some would have it) who is thinking all those thoughts. That is the real us. I repeat this here, because it is a rather hard point to grasp.

Some might ask, what's the difference? If we behave as if our thoughts are us, if we perceive the sum total of our thoughts as being our consciousness, then is there really a difference between the thinker of the thoughts and the thoughts one thinks? If perception is reality, are they not then one and the same thing for all practical purposes?

Mr. Chopra seems to believe there is a fundamental difference between the two, as do many others. I tend to agree with him/them. For instance, just because a computer can compute, can come up with answers we need, the computer is not the sum total of its computations, but is an actual thing in and of itself, an electronic machine that makes those computations. Destroying the computational results would not destroy the computer. Moreover, nobody would confuse a computer as just being the sum total of what it computes. I believe this holds true as being the same with human consciousness versus the physical brain. But that is just my personal viewpoint. This book is about you forming your own opinions on the matter.

In any case, whether one wishes to define this "thinker" of thoughts as the soul or consciousness is a moot point. The main point here is that Deepak Chopra chooses to separate our thoughts out from ourselves, as actual conscious beings. He, more or less, views the true "us", as the consciousness that thinks, and the thoughts it thinks as just a product of the "thinker" of those thoughts. That, in his estimation, is the true inner self, or as some people would consider it, our consciousness or soul. Now let's move on to the next approach to consciousness:

2. Materialists. Consciousness Just a Physical Result of a Highly-developed Brain? Other researchers feel that the human consciousness is just a direct result of our physical brains. Build a better computer, in other words, and it will be capable of more complex computations. Such scientists feel that if there are enough neurons and neuron connections, if the brain becomes complex enough, that human consciousness is the natural result, the compilation of all the inner workings of the physical brain. Consciousness, then, is just a natural byproduct or result of the physical workings and interactions of various brain cells.

Such researchers perceive consciousness as just being the manifestation of a very complex physical creation—the brain, rather like the old mechanical automatons that could run on a clockwork system of gears and springs, ones that were once so popular in the last centuries. In one case, there was an automaton so complex that it could draw four different pictures. Does this mean it was conscious and capable of creative thought and could somehow exercise this creativity? Had we created a conscious being in such a way?

No, of course not. This art ability was simply the product of having a complex enough machine made of gears for this to be able to be accomplished. There was no sense of self-awareness, of free will, or any capability or capacity whatsoever for such a device to contemplate its own existence or to be truly creative. The drawings were just the byproduct of a machine and nothing more.

Those same researchers also point out that animals have a less-developed consciousness, because their brains in relation to their body size are smaller. Therefore, they are "less aware." In other words, such animals haven't achieved the necessary complexity in their brains to have true consciousness as humans do. As a consequence, whether they are less aware, or perhaps not really aware at all would seem to be what these scientists argue.

Other scientists disagree. They point out that probably the one great difference between something that is conscious and something that is not, is freedom of choice, free will. Animals seem to have that as much as we do. It might take them longer at times to exercise it (ever see a cat sitting and acting as if it was trying to decide something before it suddenly jumps up and runs out of the room, or just stays there, instead, for example?), but they do seem to have free will.

Even as we do. And having the freedom of choice, as in the choice to come and go, to eat or not, to lie down or stand, or whatever, they must have a consciousness and a fairly well-developed one at that. Would we argue that a severely, mentally-impaired human is without any consciousness or is in no way self-aware? Probably not, at least to date as I've seen no evidence for that anywhere, thankfully.

So some researchers say that no matter how complex our computers seem to get, even our supercomputers or quantum computers, they do not manifest self-awareness or freedom of choice, and therefore, probably are not conscious, despite their complexity. And it then follows (and they also argue this), that the same holds true for the human brain. No matter how complex the brain, there is no natural result for such brains creating a separate conscious entity but rather consciousness is just the manifestation of a complex brain.

However, if this is so, one could ask if a supercomputer is conscious, aware of its own existence? It would seem not, at least not so far. So although our brains are complex, some researchers even argue that human consciousness may not be just a "simple result" of that complexity, but rather is just an illusion of sorts, or so such researchers' arguments go. Now for the next idea concerning consciousness:

3. Dual Nature of Consciousness (Dualists). This is the belief or theory that consciousness is a product of both "a soul" and the physical brain, that the two interact. This, they think (as the last of the three major approaches as to what is the origin of the human consciousness) might be the result of a quantum nature to the brain. Some scientists point out that there has been the discovery of very tiny physical structures in the human brain that are on the Nano (sub microscopic) level. These structures are called microtubules. These are so small, that the researchers argue they can function on the quantum level, and interact in states of superimposition, therefore. All subatomic particles seem to have this ability. They can be in more than one state at a time, and as counterintuitive as this may be, it is a fundamental precept of quantum physics, this "dual nature" of subatomic particles. Therefore, they say that the human mind may be of a quantum nature. Remember Descartes thinking human minds were dual in nature? This was before the idea of quantum physics became generally known. So either Descartes was a genius (which he may well have been) or he was singularly lucky to come up with idea of the "duality" of the human mind.

However, opponents of this idea argue that the human brain is not a subatomic particle and instead exists in the "macro" world, the world of everyday things like, trees, cars, houses, and people. So the human brain cannot have varying states of simultaneous existence, or have superimposition as subatomic particles can.

They also point out that quantum computers, in order to function, must be a super-cooled temperature in order to work, and the human brain is hardly in that category, because quite warm and also "too wet," for superimposition and quantum effects to occur. Yet, other researchers also counter this by saying that we can't know for certain if this is a drawback, that the human brain may have found a way around this problem even though we humans haven't. After all, we can create computers but not nearly as complex as the human brain is, at least not yet. So the jury is still out on that issue.

There is one more thing to consider here and that is the result of the complexity of the human brain creating something that is really rather outside of three dimensions. An article at *Sciencealert.com* in April of 2018 stated this:

> "Last year, neuroscientists used a classic branch of maths in a totally new way to peer into the structure of our brains. What they discovered is that the brain is full of multi-dimensional geometrical structures operating in as many as 11 dimensions."

The researchers found that human brain neurons could organize themselves in networks that they call "cliques," and that these cliques can organize themselves in structures that are, as they put it, a:

> "...high-dimensional geometric object (a mathematical dimensional concept, not a space-time one) ...."

They go on to say that even in the tiniest bit of the brain, there are "tens of millions" of these and that they can organize in up to 7 dimensions, or sometimes can go even higher, up to as high as 11 dimensions. The consequences of the brain having this capability are largely unknown, but this ability might be how the physical human brain gives rise to consciousness.

A World Divided. Let's just quickly reiterate the three approaches as to what consciousness is, it's origins. There are three different approaches as to what consciousness might be:

1. The Idealist approach, or one who believes the consciousness is totally separated from the body, is an entity independent of any physical nature, such as our brain.

2. The Dualist, who believes the consciousness is more than the result of the physical brain, is more than just a natural consequence of the laws of physics, but also, it interacts with the human brain, as the brain does with it. Thus, there is a "dual" aspect to consciousness, both physical and nonphysical

3. The Materialist or those who believe in Physicalism. They claim that consciousness is strictly the result of normal laws of physics, that consciousness simply arises when one has evolved or developed a complex enough brain to allow for it to happen. Given an extensive and complex neural network or net, consciousness will be a side effect or natural consequence of such a thing

As you can see, the modern world seems divided on the concept of what produces consciousness. The "physicalists," as in many doctors, researchers, and scientists say, that consciousness is not separate from the body but rather is merely a function of the brain.

Physicalists point to the fact that when the brain is damaged, this can cause behavioral changes in the person involved. The victims of such damage can become violent, suicidal, uncontrollable, undergo dramatic personality changes and/or have suddenly different morals, or any number of changes occur. They point out that people born with physical deformations in the brain often are mentally handicapped, sometimes to the point of being barely functional, and in some cases this is so severe, that they seem not even to be self-aware or if so, just barely so.

Others, because of brain damage or brain birth defects suffer from being savants or even acquire genius capabilities. This is an odd phenomenon because these people, though often suffering strong mental disabilities can perform mental feats that border on the actual miraculous. As Wikipedia puts it:

"One in a million people [suffers from being a savant]. Savant syndrome is a rare condition in which someone with significant mental disabilities demonstrates certain abilities far in excess of average. The skills that savants excel at are generally related to memory. This may include rapid calculation, artistic ability, map making, or musical ability ..."
[Clarification added.]

As to how or why this is, medical researchers are unsure, other than to theorize, but such theories are based on very little actual hard evidence. To date, it isn't known why some people are savants and others, equally mentally handicapped, are not. Moreover, no one seems to understand how they can accomplish what they do, when even very intelligent people, those with very high I.Q.s, are not capable of doing the same.

Even so, physicalists, those who believe all things about human consciousness are determined by material functions of the brain, point to all these physical problems of the brain and how they affect a person's personality, ability to function, and behavior as proof that the human consciousness, or soul, if you will, is just a function of the physical brain. Damage the brain, and the person's consciousness suffers.

But does it? As idealists or "spiritualists point out (and not the spiritualists as in psychic mediums, but instead those who believe the spirit or soul is a separate entity from the body), a brain is just a physical object, even as a music CD or DVD is just a physical object. You can destroy the CD (or DVD), thus damaging the thing so it won't play the music, but that doesn't destroy the symphony that created the music recorded on CD/DVD. Nor does it destroy the player. The damaged disk is merely made inaccessible to the player, so the music can't be recovered and played.

The same with a computer. You can damage a computer to where it won't function, but that doesn't destroy the formulas that have been run on it. The computer is a physical thing that performs calculations but it is not the calculations themselves. It is merely an interface that allows the user, you, to have those calculations performed. Destroy the medium, and the ability for you to perform those calculations on the computer (or brain) stops, but it doesn't stop the fact that the calculations are separate from the computer and that you are no longer you even if your brain has been damaged.

Therefore, idealists or spiritualists think of the physical brain as being this way, as an "interface," just as a computer is an interface. Again, for them, a computer merely runs the programs or apps loaded onto it, but it is not those programs and neither is it the programmer.

For instance, if you use a computer (cellphone) to talk to other people, via messenger, and/or texting of some sort, damaging the phone or computer only damages those physical items themselves. It doesn't damage the other person with whom you were communicating with or who was communicating with you. Both of you are still just fine. It is just the interface that is damaged. The same with a symphony. Just because the CD/DVD may be damaged and won't play the music doesn't mean the symphony that created that music is affected or ceases to be. It simply can't be played on that particular CD, memory stick, or whatever.

Idealists believe, in essence, there is a "thinker of thoughts," and that the thinker is not just the thoughts themselves, but that behind those thoughts there is a "someone" thinking them. Some have a variation on this, and feel that thoughts come from somewhere else and our brains are just receivers of those. Rather like a radio show being broadcast and a radio is just acting as the receiver of that broadcast but is not the actual broadcast itself.

Even so, even if thoughts are being broadcast to us, someone or some "thing" is thinking those thoughts and each person thinks of different things, has different ideas, and notions. So if thoughts were being broadcast to our brains, and we are just receivers, how is it everyone has a different set of such thoughts, beliefs, and behavior patterns as a result? We would all seem to be very unique in that regard.

So people who believe that consciousness is separate from the body point out that a damaged receiver, the brain in this case, means that the messages of the broadcast just aren't getting through correctly. Just as a television may lose its picture but still have sound, a damaged human brain might get part of the consciousness coming through to it but not all, and so of course, the person would then malfunction, have behavioral problems and issues.

A badly damaged brain would result in severe mental issues because it is a badly damaged interface, even as a computer is, but it is not the consciousness itself that is damaged, any more than a computer is the formulas themselves.

Idealists, or Spiritualists believe the consciousness is still there, just unable to manifest itself sometimes because of a badly operating brain. Dualists also incorporate this idea to some degree, as well. They theorize that there is extensive interaction between the physical brain and the mind, and vice versa, but that the consciousness is not entirely dependent on the physical brain, either.

Personally, I rather like this idea, because it explains a lot. The consciousness remains intact but it simply can't "get through" because of the medium, the brain it is relying on to do so is damaged. This might explain why some people have out of body experiences (OBEs), and near death experiences (NDEs). The consciousness in such cases is either partly already outside, or moving "outside" of the body under such circumstances. Moreover, it means the physicalists are right as far as they go, but it also means the idealists, those who believe in a soul or separate consciousness are right, as well. A happy outcome for everyone, as it were.

This is something that I do feel to be innately true and this last is just a personal observation. I am "aware" that I think thoughts. I am aware of the fact that I am aware. My thoughts are not me, but rather I am the thinker of those thoughts. And if the consciousness is truly a separate entity, perhaps either originally created by the brain but then becoming an entity in its own right, or wholly separate from the brain, then that explains just about everything one can imagine about all the things with which people struggle in that regard.

As just an example, it would explain reincarnation, the soul or consciousness moving from one body to the next. It could also explain remote viewing, clairvoyance, telepathy, and other psychic instances. A separate consciousness would also act as an explanation for near death experiences and out of body experiences.

Consciousness being intrinsically separate from the body even could explain ghosts and other apparitions. Such things as automatic writing (which some famous authors say is how they wrote their works), claim that someone else guided their hands in the writing, could be explained by this. And before we discount automatic writing, remember, some very famous authors claim that what they wrote wasn't their work, but rather they were guided to write it, as with the example of Harriet Beecher Stowe:

"Harriet Beecher Stowe, the author of Uncle Tom's Cabin, claimed that she did not write it: it was given to her;" "...it passed before her." – **[see References for link to source]**

And in case you think this is a one-off instance, think again. Many authors have claimed their hand was "guided" by something other than themselves. Even the Old Testament of the Bible had parts written in such a way, and this was claimed by the writers, themselves, as in:

"2 Chronicles 21:12: "And there came a writing to him from Elijah the prophet saying ..."

Chapter Conclusion. We also see that we can describe consciousness, as in "being cognizant," but we don't really "know" what consciousness is. We can describe the state of consciousness, or we can describe the lack of it, as in someone is aware, or unaware of their surroundings. They are either conscious or unconscious, wakeful or comatose, or perhaps even dead. So does consciousness end with death?

Many people "believe" that it does, but can we be sure? Or is it just a vain hope on our part that somehow, we might survive our own deaths. Well, the answer to that remains to be seen. However, there is tantalizing evidence, scientific evidence even, that consciousness might survive even death of the physical body. Perhaps, consciousness might even be immortal in some fashion or fashions. And once more, perhaps without consciousness, the universe might not even be able to exist. Moreover, perhaps we even create reality as we go along, and it takes a consciousness to do that. The science of quantum physics has led to some very interesting conclusions and intriguing ideas about all this of late. We will discuss these subjects in depth later on in this book.

One final thought here; as one scientist pointed out, the fact that even if any of these means of creating the human consciousness from the physical brain hold to be true, and they do create the human consciousness, THAT STILL DOES NOT EXPLAIN WHAT THE HUMAN CONSCIOUSNESS IS!

No matter how our consciousness may have been created, it doesn't explain what the nature of consciousness itself is. How is that we supposedly have free will (if the universe is not deterministic as some believe)? How is it we can have thoughts, ideas, imaginings, ponder the nature of the universe and ourselves, be aware of our surroundings and even of our own existence? How is it that we can even question the meaning of our existence?

No scientist or researcher has even come close to answering that question yet, or even can come up with a rudimentary theory. Human consciousness, whatever its origin, seems to embody not only self-awareness and free will, but the ability to contemplate its own existence and that of the nature of reality. And more, it can research, investigate and test such theories to see if they are true. An incredible thing in and of itself, is it not?

Now that we've discussed the nature of consciousness and its possible origins and what it is, it's time to move on. Even if we are sure of how consciousness arises (and we aren't), we still want to know if it can survive the physical death of the human body. And there is another question I have to ask at this point before we continue: If the human consciousness is a product of the physical brain only, what happens to it when the body dies? Immediately after death, as one scientist put it, all the physical mass, the energy of the body is still there, but why is it no longer conscious?

If nothing is missing, if all is still there, shouldn't the consciousness still be there also? And the energy can no more just instantly vanish than the body can. That would be in violation of the First Law of Thermodynamics. So why is the person dead? What's changed? Where have "they," that thing that makes them, "them," gone? Again, what has changed?

Something must have, for the consciousness seems to have vanished, gone, or ceased. Yet, sometimes an individual can be revived, even after long minutes, and the consciousness returns. How is this also possible? When is dead, really dead? Keep these questions in mind for later on in the book.

Chapter 2—Consciousness Overall

"Idris: Are all people like this?
The Doctor: Like what?
Idris: So much bigger on the inside."
—Neil Gaiman

    This is probably one of the most in-depth chapters of this book, because the topic is a truly profound one. People have argued over this subject, the idea of consciousness and the universe forever and nobody is sure if we are anywhere nearer any solid conclusions about it.

    What is consciousness? What does it mean to be conscious? Does consciousness even exist, or is it a mere illusion foisted upon us by our own bodies to make us think we are in charge when we are not?

    For example, a number of studies now show that many if not most of our so-called "conscious" decisions are actually made subconsciously and it is only "after the fact" that we think we made such decisions consciously. If you reach for a cup of coffee for instance, your subconscious has already decided to do this and so it is just an illusion that your conscious mind made the decision when, in fact, it did not.

    Yet, even if many or most of our decisions are made at the subconscious level, we still do think. We still wonder. We still question. And whether consciously or subconsciously done or not, we still choose at some level. We all still investigate and seek answers. Moreover, we still form opinions about things.

    This then, is what we might choose to think of as our conscious selves, or our consciousness. As René Descartes, a noted French philosopher once said, *"cogito, ergo sum,"* which is Latin for, *"I think, therefore I am."* And that thinking, those feelings we have, combine to form our consciousness.

    To put it more officially, as the online dictionary, Dictionary.com puts it:

"...consciousness, the state of being awake and aware of one's surroundings. **[as in]** *"she failed to regain consciousness and died two days later."*

[or]*...awareness, wakefulness, alertness, responsiveness, sentience..."*
[Clarification added.]

    Moreover, although much that we do may indeed be at the subconscious level and we are unconscious of this fact at the conscious level (gets convoluted, doesn't it?), that doesn't mean that we don't actually think, don't search for answers and try to comprehend the universe around us, as well as ourselves. Even if our subconscious is making most of the decisions for us, our conscious mind still seems to be there, awake, aware, and alert (at least some of the time…).

So does consciousness actually exist? Yes, it seems that it does, although again, many of the decisions we ultimately do may not be at the conscious level, and we only think that's the case. Our consciousness also seems to have the ability to "believe." Sometimes and as we have seen, this ability is so strong that it can cause physical alterations in our health and well-being.

It may also do more. Consciousness, as being a conscious observer may be a necessary requisite in the universe. Indeed, some researchers think the universe couldn't exist without it. This is especially true of the science of quantum physics, where innumerably repeated experiments seem to show that this must be the case.

We know that to be conscious is to be aware, and in our case as humans, self-aware. However, for only about a century now have we realized that consciousness may be absolutely necessary for there to be something of which we can be aware of, as in reality around us. The old idea that does a tree falling in the forest make a sound if nobody is there to hear it does, in fact, seem to true in a way, if one is to believe quantum physics. We will get further into that a little later on this book.

Despite being aware, science has a very odd time explaining how the human mind can develop "qualia," or as one article on consciousness puts it:

"…it remains an open question how humans develop self-consciousness and obtain basic knowledge of the type called qualia (Chalmers, 1995). The hard problem of consciousness is the problem of explaining how and why we have qualia or phenomenal experiences and how sensations acquire characteristics, such as colors and tastes." -- Consciousness in the Universe is Tuned by a Musical Master Code, Part 2: The Hard Problem in Consciousness Studies Revisited. -- Dirk K. F. Meijer, Igor Jerman, Alexey V. Melkikh and Valeriy I. Sbitnev.

There is another aspect to consciousness, as well. Many scientists believe that consciousness is not just limited to humans. They point to the fact that many animals have the basic brain and neurological capabilities similar enough to us in nature that they, too, may be conscious, as in being aware of their surroundings. As **Wikipedia** points out in its discussion of consciousness:

"…Consequently, the weight of evidence indicates that humans are not unique in possessing the neurological substrates that generate

consciousness. Non-human animals, including all mammals and birds, and many other creatures, including octopuses, also possess these neurological substrates...."

To paraphrase one researcher, if a cat, dog, horse, or other animal can experience joy, for instance, as in loving to be scratched or petted, then they must be conscious, at least to some degree, because they are aware of their surroundings. And that is one of the main definitions of consciousness, awareness of one's surroundings.

Indeed, their consciousness may be limited compared to ours, even as some people who are badly brain damaged in some accident might be so limited, but they are still definitely aware. Therefore, they are conscious. If not, they are unconscious as in being comatose. Even then, many people argue the person's consciousness is still intact and may actually still be functioning. When we sleep, our mind dreams, even if we don't remember having dreamt upon awakening. Over 60 percent of people who have undergone anesthesia, later say they had "things" equivalent to dreams. Others do not remember having these, but as with sleeping, it doesn't mean that just because they don't remember having them that it hasn't happened.

Moreover, we probably should not get too hung up on the idea of the conscious versus the subconscious because many researchers feel it is just that part of our makeup that we aren't focusing on, but which is still active. Ever drive down a road while deep in thought? You pass through several stoplights, and it is only after you do this that you can't remember if they had been green or red? Yet, your mind still managed to navigate you safely through those stoplights. That's the subconscious operating.

Your subconscious took over and acted to guide you safely through all those traffic lights (and no, you didn't run any red lights even though you don't consciously remember because your subconscious took care of that). Because your conscious mind was focused elsewhere, the subconscious took over handling the more mundane task of driving as it does with so many things, like riding a bicycle, walking down a street, etc.

Why do our minds do this? Because we can't be conscious of everything all at once around us and all the time. Contrary to Sir Arthur Conan Doyle's fictional super sleuth, Sherlock Holmes, humans aren't made that way. Why? Because there is simply too much data pouring in all the time for our brains to handle and still function and be able to perform tasks. Information pours into us in huge amounts every second from all our senses.

As **Britannica.com** mentions:

"...the human body sends 11 million bits per second to the brain for processing, yet the conscious mind seems to be able to process only 50 bits per second."

Some more refined measurements suggest that we can process about 60 bits per second, but that is NOTHING compared to the inflow of 11 million bits per second with which our brains have to contend!

So the subconscious intercedes for us and takes over in many instances, acts as a filtering system, so that what might be important can still get through to us. Of course, there is more than just that to the subconscious, but this is sufficient for our purposes here. Books have been written in countless numbers on the topic of the subconscious.

Also, it is important to remember that even though our subconscious may guide us safely through traffic lights, it is not infallible. Sometimes, it can get mixed up because it relies heavily on habit, things learned from past and similar events. Ever reach for a light switch in a room when your mind is really focused on something else only to belatedly realize that you are reaching for the switch on the wrong side of the doorway, and then only realizing this after you have fumbled to find it and it isn't there? That's because the subconscious relied on past events, ones of a similar nature to try to meet your current needs. Only when this fails to perform well do you become consciously aware of the fact and have to then consciously intercede to correct the situation.

By the way, despite all this information on consciousness, scientists really don't know what it is. They describe how it behaves, as in "wakefulness, awareness," etc., but that doesn't really tell us what consciousness intrinsically is.

For example, if you have ever had a true epiphany, knew in an instant that something is fundamentally true, sublime, or whatever, and then tried to explain this to someone else, they will often give you an odd look of disbelief, or simply can't comprehend the true nature of what you had experienced. It is rather like trying to describe or explain colors to a person that has been blind all their lives. One has to actually experience color at some point in their lives to be able to understand truly what a color is.

As another example, is the question of dreams. Several researchers state that it seems likely dreams are the result of a consciousness (even if in a subconscious mode) creating its own space-time continuums for however a short of a time they might last. In dreams, entire regions, places become real for the dreamer, complete with color, sound, touch, smells and even sex. Although dreams are often disjointed and constantly changing, they are very real for the person having them. In fact, while dreaming, most dreamers cannot discern the distance between dreams and the "real" waking world.

Moreover, not only is the dreamer in the actual dream, but simultaneously, he/she is also the dreamer of the whole dream. The creator of it all, including all the characters in the dream, the surroundings, etc. Inside one's head, then, realities are created in which the "creator" (the dreamer) not only makes these space times of whatever size, shape, and description, complete with smells, colors, textures, and even sex, but also the "creator" is actively a participant in the dream itself.

How is the possible? Again, for the dreamer experiencing the dream, that is the "real" universe, and not the one his or herself is awake and conscious in most of the time, but no less real for the dreamer while dreaming for all of that. Ever wake up from a dream and marvel at how "real" it seemed to you? For all intents and purposes then, that dream was real for the dreamer while dreaming it. People have even suffered heart attacks, strokes, and other repercussions from their dreams, especially when they are nightmares, so the effects of dreams can certainly be real enough, including even resulting in death. So which universe is our real one? Most would argue it is our every day one, but then how to explain the very real nature of many dreams and their physical consequences? How do we integrate all that into the idea of what consciousness is?

Another example? There is the idea or concept/thought experiment of the Boltzman Brain. According to the laws of probability, it is even more likely that a brain might spontaneously come into existence in space somewhere, rather than to evolve on a planet over billions of years. Regardless, though, of which is the more likely, the percentage chance of this happening, although very small is not zero.

This means given enough time and the right circumstances, such a Boltzman Brain might spontaneously arise somewhere in space by pure chance of the right things coming together to form it. The brain would come into existence complete with memories of a life, a world with people and trees, etc. It would think it had lived all along in this made up world, and could "live" in this world just as people "live" in their dreams. Without any outside sensory capabilities because it would be just the brain and nothing more, the mind of the Boltzman Brain would only have its own imagination to rely on to create such a reality.

The amazing part about this thought experiment is that, theoretically, it should already have happened, and possibly twice, given the age of our universe. This means that all this around us; the world, the stars, trees, houses, cars, us, and others, might just be a figment in the mind of a Boltzman Brain, one living inside its own imagination, since again, it would have no exterior input from the universe around it to tell it otherwise.

So, it is possible we are all just part of the fevered dream of perhaps some insane and disembodied brain floating in space!

This was a real thought experiment and one based on the laws of probability. Ludwig Boltzman described the brain has spontaneously arising, complete with memories of a life and everything else, ones not discernible from reality for the brain. So the question of what constitutes consciousness and what constitutes reality seems to be a very fluid and vague question, one with multiple possible answers. Of course, as with all such controversial ideas, there researchers who oppose the possibility of it, arguing about the basic premise of the thought experiment being wrong.

There are people who have at times felt "truly one with the universe." When asked to describe this feeling to others they can go on at length, but there is still a problem. The description of something is simply not the same thing as experiencing the real event for one's self. You might get an inkling or sort of an idea or understanding, but that is not the same thing as living the moment yourself, having that feeling of "oneness."

Again, it is rather like trying to describe the color blue to someone who has been blind all their lives. It simply can't be done. You can use terms like color, shades, tints, and even give examples of things that are blue, like the sky, but that holds very little meaning for someone who has never seen any colors or shades at all!

Consequently, although we can describe the attributes of consciousness, the "symptoms" of it, if you will, we still don't know what consciousness really is. What allows us to be aware? What allows us to be "wakeful?" What allows us to create realities indistinguishable for this one in our dreams (at least, until we wake and realize how unlikely they were)? How is it we can think, feel, emote, imagine, and ponder?

Yes, again, we can describe it, even as we can describe a train roaring by us, but that doesn't tell us how the train actually does that. Indigenous Americans once referred to trains as the "iron horse." It is all they had to equate it to. At the time, they had no concept of steam power, gears, boilers, and such. So although they could sort of describe a train, they didn't really know what it was or how it worked, other than that it did. Now of course, we all do understand the basic concept of steam power.

Concepts of consciousness vary as to what it is, descriptions of it, and how it has come to be. Some researchers argue that consciousness is really just an illusion, a byproduct of our physical brains. Other researchers believe there are different levels of consciousness. This last includes scientists, religious persons, as well as many of those who believe in the metaphysical.

In fact, some of the ancient religions have argued much the same as some modern researchers are starting to do now. The Hindu religion has long held that all (everything) is illusion, even that of consciousness. Some researchers even go so far as to say that we are merely constructs in some vast simulation, as conscious players might be in some computer game. Still others argue that consciousness is universal and that all things are conscious to some degree, even rocks, trees, stars, and planets. There are even those who say the universe itself is conscious. Some of these concepts we will discuss in more detail later on.

Chapter Conclusion. As of now, we still really don't know what is consciousness. However, there are many researchers in many fields trying to figure out just what consciousness is and how it arises. Are our minds, our souls just a byproduct of the physical brain or is the total greater than the sum of the parts? In the next chapter, we will discuss the possible origins of human consciousness and what it all might entail. One thing to remember, though, and that is what we view as reality is merely a recreation of what we see, touch, smell, etc., all around us, and that recreation in our minds can be flawed and sometimes very misleading. For instance, check out this "optical illusion." Please note, this is an illustration in color and may not work in black and white print versions of this book):

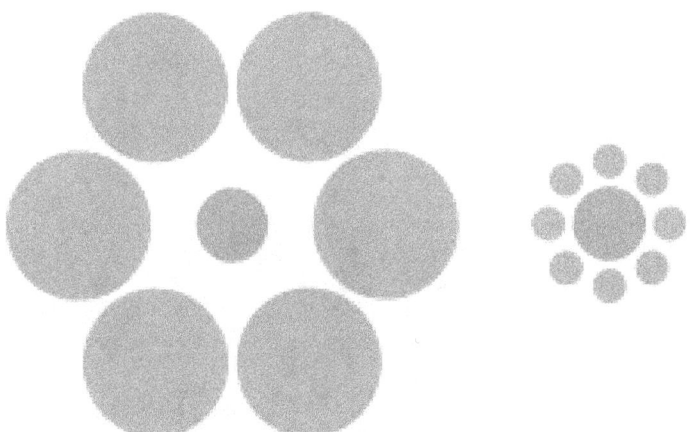

Source: Wikipedia

In this illusion, the mind is convinced as "reality" that the orange circle in the center on the right is bigger than the one on the left. In "reality" it is the same size but the recreated image in our mind insists the one on the right is bigger. Or this black and white illusion:

Source: Wikipedia

The square on the left is the exact same shade as the square on the right but our minds insist it is darker. The recreation of what we are seeing in our brain is not the actual reality.
   (https://commons.wikimedia.org/w/index.php?curid=75000950)

By Original: Edward H. Adelson, vectorized by Pbroks13. - Own work, CC BY-SA 4.0, *https://commons.wikimedia.org/w/index.php?curid=75000950 By Original: Edward H. Adelson, vectorized by Pbroks13. - Own work, CC BY-SA 4.0*

In the above optical illusion, Square A and B are actually the same shade, but our minds cannot perceive it as such. The mental recreation insists on showing them as different shades for us. Therefore, our minds simply are not recreating actual reality as it should appear, but instead gives us the illusion instead.

Chapter 3—The Power Of Belief

"Existence is just a myth we all believe in, while it's just a mere blink of an eye."

—Meeran W. Malik

Perhaps, before we get into the idea of possible immortality, we should discuss the idea of "beliefs." After all, they play a very powerful part in our lives, as in our "belief" in an afterlife, and one has to have them to belong to any religion on Earth. Different religions may believe different things, but they all have a set of "beliefs."

Moreover, there seems to be evidence that we may be "hardwired" for having beliefs. So why do we have them? Do beliefs hold any real, intrinsic power? Do they help us? Or do they hold us back and even harm us? Are beliefs just that and nothing more, having no power at all, any more than a child's wish might?

The main definitions of belief, according to the Merriam-Webster Dictionary is as follows:

> *1: a state or habit of mind in which trust or confidence is placed in some person or thing **[as in]** her belief in God a belief in democracy **[or, as in]** I bought the table in the belief that it was an antique. "contrary to popular belief."*
>
> *2: something that is accepted, considered to be true, or held as an opinion: something believed; an individual's religious or political beliefs especially: a tenet or body of tenets held by a group **[as in]** the beliefs of the Catholic Church.*
>
> *3: conviction of the truth of some statement or the reality of some being or phenomenon especially when based on examination of evidence **[as in]** belief in the validity of scientific statements."*
> **[Clarifications added in brackets.]**

To put it in simpler terms, and as one other definition would have it, *"belief is having faith, a faith in that which cannot be proved by scientific or any ordinary means of obtaining such proofs."* Furthermore, and as yet another way of putting it, belief is having faith in something for which there is no [real] direct evidence of a factual or scientific nature upon which to base such a belief.

To make matters even more difficult, often and conversely having faith in something is to "believe" in that which cannot be empirically proven at all. That is, faith is having a strong, perhaps unshakeable conviction in something, but it is something that cannot be supported by any objective evidence. Such a conviction or belief cannot be verified or tested by the observations of others, and often would appear counter to logic and direct, scientific experience.

Our problem here is that the definition of beliefs and faith form a closed loop. The definition of faith is to have a belief or beliefs in something. A belief is to have faith in something which cannot be proven as true or real.

It's a bit like asking a person why they believe in God, and they answer because the Bible tells them so. When you ask them why they believe in the Bible, they will often answer because it is the word of God. Again, a closed loop. Their "evidence" for God is the Bible, and they believe in the Bible as such evidence because it is "the word of God."

Now, my purpose of defining belief and/or faith here is not in any way to belittle or demean the ideas of anyone's beliefs and faith. Whether one chooses to "believe" or "have faith" in the idea there is a God, is not the question here, and most decidedly, not the point. Rather, the point here is that belief and/or faith, are convictions that are usually of a highly personal nature and are unprovable by using the scientific method. For instance, it is said that no two people believe in God in precisely the same way. Each person has their own take, their own unique perspective or slight variation on the idea of an ultimate supreme being, and this is so even if they belong to the same church, mosque, temple, or whatever.

As we all know, there are many beliefs about God and various religions in the world, and some are very different in what they believe from others, even as to whether there is a God in the traditional sense of the word at all. Some believe in multiple gods, while other religions have a very different idea of just what the concept of God is. Pagans believe in multiple gods for instance, while perhaps other pagans believe just in nature, as in there being a "Gaia." Still others believe in a powerful creator but one who does not answer individuals' prayers, while yet others believe in a personal god who does exactly that.

Beliefs in all these different things, beliefs held by the believers of such religions or "isms," can be very strong, terribly strong in some cases, and this has been so throughout history, it seems. People fight, quarrel, battle, and die for their beliefs on occasions, or even kill others who disbelieve.

Therefore, in the sense of a person's strength in their beliefs, there is a power attached to them. Beliefs, real or not, motivate people. There is the power to do good, as to live a saintly life and help others. There is also the opposite, the power to wage war, to torture, and kill those who do not believe what you might believe. So yes, there is a mundane power in beliefs, one that can cause people to behave differently about themselves and toward others.

However, this is not the type of power of belief that I am talking about here. My question is more simple; does belief in something give the believer actual power in a supernatural sense? If one believes devoutly in real magic, can one make real magic happen, as some followers of such groups as Wiccans or other types of pagans believe? Can they actually perform magic if they devoutly believe that true magic is a real thing, as in the Harry Potter type of magic?

Before you laugh at this idea, think about how the same questions also apply to religions, something many people do devoutly believe in. For instance, if one is a true believer and lives a saintly life, can one perform miracles, as the Catholic Church and other churches truly believe is possible? Can there be virgin births, for example, if the power of God is invoked, as with Mary of the New Testament? Can one reach a state of Nirvana, and transcend the physical if one mediates enough as Hindus believe? Can one achieve miracles by living in real harmony with the universe, or by requesting God to perform such miracles for them, as Moses supposedly did?

All religions seem to rely on this supernatural or paranormal aspect of things, that there "is something more" than what we see around us. That there is "more to the universe" than we know or comprehend as the character, Horatio, was told in Shakespeare's play, *Hamlet*.

Again, my purpose here isn't to challenge these beliefs or any such belief systems. The truth is; I simply don't know for a fact that any of them are true or untrue. For me, therefore, the jury is still out. For me, personally, God cannot be proven, but neither can he be disproven, so until somebody manages to do that, one way or the other, I refrain from choosing either side.

Yes, I find little scientific evidence to support many such religious beliefs in many ways, but by the same token, there does seem to be some small evidence for them, as well! Furthermore, a lack of evidence doesn't mean that something isn't true. Negative results don't disprove something, but rather, just make something seem more unlikely if consistently failing any tests that can be devised to ascertain the truth of such things.

A bit confusing? Well, let me clarify with an example: The Catholic Church believes that miracles can happen. There are people who spontaneously manifest something called "stigmata" the marks that Christ supposedly endured on his body during his crucifixion. Now, we do seem to know for a "fact" that people do manifest such "stigmata." It has been well documented in some instances and repeatedly.

They bleed from the palms of their hands. They might have the bloody marks of the crown of thorns appear spontaneously on their foreheads. They can and sometimes do develop open wounds on their feet, or in their side, etc. Moreover, there are some strange effects that go along with these, as well. For instance, they can bleed steadily for days, even weeks, sometimes permanently, and never suffer for lack of blood. Also, some respected observers report a scent of lavender or other perfume-like odor from the blood. There is more, but I think this gives us enough of an idea.

There was one very well documented case of stigmata, one observed directly by doctors and even some scientists, not to mention countless other reliable and yes, some unreliable witnesses, as well. This was the case of "Father Pio." As **Wikipedia** refers to him:

"Padre Pio, also known as Saint Pio of Pietrelcina (Italian: Pio da Pietrelcina), (25 May 1887 – 23 September 1968), was an Italian friar, priest, stigmatist and mystic, [1] now venerated as a saint in the Catholic Church. Born Francesco Forgione, he was given the name of Pius (Italian: Pio) when he joined the Order of Friars Minor Capuchin.

Padre Pio became famous for exhibiting stigmata for most of his life, thereby generating much interest and controversy. He was both beatified (1999) and canonized (2002) by Pope John Paul II.[2]"

Padre Pio—Public Domain Illustration
Notice hands are covered. This is to hide
Bandages that often appear in other photos.

Now, did the power of belief in God, or God him/herself create the stigmata, which in and of themselves seem real enough? Possibly. We simply can't know that for certain. But there are things we do know for certain:

1. The stigmata Padre Pio manifested were apparently real by most accounts of doctors and scientists and did behave strangely.

2. Oddly, no reports at all of stigmata exist in Catholic Church history prior to St. Francis of Assisi. This was in the 13th Century, and of course, the Catholic Church dates its origins all the way back to almost 2,000 years ago. So stigmata, relatively speaking, seem to be comparatively, a more recent phenomenon, and simply didn't happen prior to the 13th Century when they made their first appearance.

3. Also, oddly, the majority of stigmata cases, some 85%, have been women.

4. Some stigmata cases have been fraudulent, of course, for their always seems to be hoaxers mixed up in such events.

5. Another oddity about stigmata is that they often didn't appear where they historically should have. The wounds in the hands, for example, often appeared in the palms of the hands themselves. However, we have since learned that the nails in the crucified were normally driven into the wrists, since the small bones of the hands themselves would have pulled apart under the weight of the person being crucified and so did not suffice to hold the person to the cross. The idea that the nails were driven into the palms of the hands comes from religious paintings and crucifixes from much later times, well after the time of the crucifixion and so was, apparently, wrong. This is based on evidence garnered over the years by various archaeologists and researchers into Roman practices of those ancient times.

So what are we to make of all this? Stigmata, at least to some extent, seem to be physically real. However, they weren't always with us, but came rather late on the scene in the history of the Catholic Church. Moreover, some of the stigmata markings seem to have manifested in the wrong areas of the victims' bodies. Yet, the stigmata themselves, at least in some cases, again, seem to have been real!

How do we account for this? Would God have manifested stigmata in the wrong locations on the human body, even though God is supposedly all knowing and so would have known better than to do this? Why would He/She have only started manifesting stigmata in the 13th Century but not before?

For those who believe such marks are the direct work of God, these questions have a simple answer: "One has to have faith." That's what I was taught in Catholic School as a child. Yet for me, the questions persisted. How does one reconcile the somewhat contradictory facts involved in stigmata?

Then there is the power of prayer. Multitudes believe in the power of prayer, the benefits of it. The concept is so enshrined in our everyday lives that when someone suffers a tragedy, it is almost certain to be followed by a statement from someone or other that "our thoughts and prayers are with you." Many, many politicians say this. Yet, in the case of mass shootings and such, the power of prayer seems to be markedly lacking in any beneficial outcome. All the prayers of all the people in the world don't seem to undo the outcome. The dead, in other words, stay dead. What's more, there are more mass shootings than ever before in our history, and this despite all those countless prayers.

There was a study done on prayer and what benefits it might have in hospitals for those who were very ill. The study focused on people who believed devoutly in their religions and had support groups of people who prayed for them. The other group were people who professed no real belief in God or an afterlife and so didn't bother to pray or have people pray for them.

After the study concluded, two things became apparent:

1. The people who believed in God and an afterlife died at the same rate as others in the general population of the hospital. There seemed no significant difference as to mortality rates, on a whole, for the hospitals.

2. However, the people who believed in God and an afterlife, and who were prayed over, actually had a very slightly higher death (mortality) rate than those who didn't believe in such things. This number did not amount to much more

The study's conclusions showed that prayer by groups of people and individuals seemed to have no difference on the outcome of who lived and who died in hospitals except that those who were devout might have been slightly more likely to die. The researchers theorized that perhaps the people who didn't believe in an afterlife fared slightly better because not believing there was a heaven to go to, an afterlife, made them struggle harder to stay alive than those who thought they would enter a paradise upon dying. This explanation can't be certain, but rather is just a theory as why the death rate was slightly higher for believers over than nonbelievers.

Placebo Effect. Before we depart the subject of beliefs, there is one more thing to consider and that is the Placebo Effect. Doctors, researchers, and scientists have long known that if a patient or person believes a pill or medicine to be efficacious, it often is. For this reason, over the years, doctors have often prescribed sugar pills as a form of medication. They have no healing power whatsoever in and of themselves, but because the patients involved think they do, they often find relief from their problems, at least to some degree. Other doctors, notably some dentists, have used hypnosis to alleviate pain and suffering rather than any physical medication at all.

The thing here is that because some have a "belief" something might be of help, it often can be effective, at least to some extent. No, placebos won't cure cancer, but they have been known to help with migraines, standard headaches, and other forms of aches and pains, and even much else. And as far as cancer goes, there is even some evidence that a profound "change in attitude" about one's life, and their place in the cosmic scheme of things, have even caused some people to undergo spontaneous remissions of cancers, as well. This was as noted on an old episode of the television series, *60 Minutes*. So belief, it seems, under certain circumstances, can influence one's health, amount of suffering, and quality and even quantity of life.

The reverse also seems to be true. The power of Voodoo, that Caribbean belief system, seems to use the power of sheer belief to make people fall ill, and sometimes even die! Some people believe so much in Voodoo, that they literally seem capable of sort of wishing themselves to death when a curse is laid upon them. So the power of belief over the physical, and whether the belief is real or not, seems to still have real effects.

Chapter Conclusion: As we've seen in this chapter, consciousness has certain attributes, including the ability to be aware and believe in things, such as a God or gods, angels, magic, or whatever, even when there is no direct hard evidence to support such beliefs. As mentioned, people in their millions have even died for their beliefs, whether over religion, political beliefs, cultural beliefs, or whatever. That list is a long one.

We have also seen that some of those beliefs are powerful enough to interact with our physical selves for better or worse, and sometimes to a remarkable degree. Some of the results are even described as miracles. Although we all choose initially to acquire certain beliefs, whether having been taught them as young children, for example, many people then let themselves be totally controlled by such beliefs. They surrender their conscious selves to them. Beliefs have the power to change and for those people who belief them, to act on those changes, even killing others who do not believe what they believe. Beliefs have incredible power it would seem to alter human behavior and history.

Yet, this does not seem to be the case when trying to use our beliefs to change things other than ourselves or controlling others religiously, as in using prayer to prevent a death, or aid in a quick recovery of someone else, or anything on that order. Even so, within its confines and limitations, beliefs can be truly powerful in their results for an individual holding such beliefs, and literally, may mean the difference between life and death for them or those around them.

Have you ever heard the expression, "he/she willed themselves to die?" Or, "they just gave up on living?" Yes, beliefs have power and that power must ultimately derive from its source, a person's own consciousness. Perhaps this is why it doesn't affect others directly, because it is such a personal thing.

So what can we take away from all this? Well, with regard to belief, it seems to have set limits and in very special ways. People can believe in God to the point where they develop spontaneous physical wounds, stigmata, on their own bodies. Whether or not this is an act of God, I leave for others to decide, but I'm inclined, based on the available evidence, that it is the belief on the part of the person themselves that causes stigmata to manifest, and not the result of a divine intervention from God. Evidence would seem to support this conclusion since stigmata can appear in historically the wrong areas of the body (palms of hands instead of the wrists) and to date, I haven't heard of a single case of a person giving someone else stigmata.

Are the stigmata real? Yes, it would seem so, at least in some cases. However, I think this comes about because the person suffering them "believes" so strongly that they, themselves, end up by causing such markings to appear on their bodies. Again, this would explain why they can appear in the wrong place on the body because, and let's be honest, no religion believes their God makes such mistakes, let alone such obvious ones as locating such markings in the wrong place. Consequently, then, it must be the person themselves manifesting the stigmata, rather than some God.

This would seem to be a logical conclusion. It also explains why there were no stigmata prior to the 13th Century. Never having heard of such a manifestation before, even strong believers in Catholicism simply didn't (up until that time) manifest stigmata until after the first such instance of them became generally known. It wasn't until the first manifestation occurred that others followed suit, and began to bear the stigmata, as well.

As to the power of prayer; well, if the hospital study is anything to go by, and the fact that millions pray for those who died in mass killings repeatedly is any example to go by, then prayer doesn't seem to have any power at all, at least in stopping deaths. It may be a nice thing to do, may help the grieving to have people do that for their loved ones, but it doesn't stop the rate at which people die in a hospital, and it doesn't bring back the dead from mass shootings, despite all "our thoughts and prayers," notwithstanding.

The placebo effect, however, does seem to be real and has been used to medical advantage in any number of ways. To go even further, in some cases, it has resulted in astonishing "cures." Again, this seems to be an internal thing for the individuals involved. It is not the result of an outward influence of people praying for them, but rather from their own internal beliefs as to whether a placebo is real medicine, or their state of mind and sense of place in the universe results in such "miracle" cures, or relief of pain and symptoms. The placebo effect works. We simply aren't sure why, other than that a person's belief in something, placebos can have results for them, personally, if not for others.

So is their power in belief, besides the mundane version of acting on such beliefs in a physical sense by evangelizing others and/or fighting those who are "nonbelievers, etc.?" Well, there does seem to be such a power in belief, but it seems to be focused internally on the individual themselves who maintain such powerful beliefs, rather than on manifesting changes in others around them, reports of miracles notwithstanding (and which to date seems to have no real scientific evidence to support them). It seems one can create stigmata in one's self, make oneself feel better or actually get physically better, but one can't raise the dead, as it were.

In others words, no matter how strongly you might believe in magic, it seems unlikely that such a thing as "real magic" exists. It might, actually, but as of now there is no real evidence, repeatable scientific evidence that it does.

However, and again, the power of belief can make changes manifest in the human body of the believer themselves, it seems, at least in some instances. So for the individual with such beliefs, there seems to be power over their own body, at least. Whether this is a product of the belief itself, the mind or person's consciousness, or the integration of both these things that causes this effect is still unknown. Yet, it seems beliefs can physically change some things, if only about ourselves, and that seems to stem for our own consciousness.

The questions here for our purposes are; does the power of belief extend to their being an afterlife? And if belief is a product of our consciousness, how does that work exactly? Is the state of the consciousness involved the real power manifesting here? Does the power of belief or consciousness, in other words, create an afterlife for the believer involved? We will delve more deeply into this in the following chapters. Now let's turn our attention to consciousness because it is from this that all our beliefs, thoughts, and even actions stem.

One last thing, though; for some reason, the human animal seems to be predisposed to having beliefs, which other animals don't seem to be capable of having. No one is certain why we seem to be "hardwired" to believe in God or religion, at least, but science has strong indications that this may be the actual case.

Why?

Well, it may be that such hardwiring has an evolutionary positive to it in that it somehow helps us to survive. Some researchers think it might help us to maintain a healthier mental state, reduce depression and perhaps even reduce the potential for suicide in an environment that is otherwise essentially random and chaotic and so anathema to the ordered consciousness of humans. We live in what seems an often violent and meaninglessly random universe. Maybe, consciousness needs a buffer? To be conscious perhaps, is to need some stability, some sense of a higher order to offset the general injustices of life?

Therefore, consciousness might be inherently unstable without having a hardwired predisposition for beliefs in a religion of some sort. Maybe, such hardwiring acts as a balance, that a belief there is an actual higher order of existence, one that counters the apparent disorder of our lives and everyday random events of our universe.

Maybe, humans need the concept of a supreme being of some sort that incorporates an ultimate sense of justice to help us survive in a world of rank injustices. If Hitler must spend eternity in hell for his actions here on Earth, then that counterbalances the sheer horror of the terrible atrocities he's committed. If a person reincarnates as an insect or some base creature, perhaps it's as a punishment for crimes committed in a prior life?

Balance, some hope or degree of order, if only in our minds to counter such a tumultuous place as this reality seems to be might be necessary so that a consciousness doesn't go insane. A built-in safeguard or failsafe, as it were, but one that we know doesn't always work. In a cruel, uncaring, and often desperately vicious existence, one that more often than not seems to have no rhyme or reason to it, perhaps consciousness needs the illusion, at the very least if not the actual reality, of a supreme being to make sense of it all.

As a child, and even as an adult, often I had questions about how God could allow certain terrible things to happen if "He" existed. There were several answers, invariably, but all much in the same vein, as "because it is God's will," or "God works in mysterious ways," or even, "we can't know the mind of God." So even when we don't understand how a supposedly just and merciful God can allow certain horrible things to happen, many still cling to the belief that there must be an underlying reason that "makes sense," even if we mere humans can't fathom it. And who knows? Perhaps those people are right? We don't have evidence to suppose this is true, but neither do we have concrete evidence to refute such an idea.

Of course, this raises the concept of "justice," as well. Why do we perceive things as just or unjust, good or evil? Again, the source might be evolution. Snakes and spiders might be poisonous, so most people view them as bad, for example. So such perspectives of good and evil might be things that help us to survive versus those things that do not, but again, that is only a supposition by some researchers.

Nor does it explain why we have a sense of aesthetics, of beauty versus the ugly. With regard to other humans being pretty, or ugly for that matter, the cause may be evolutionary in origin, us wanting to breed with people who appear to have good genes and this is reflected in their looks, but it still doesn't explain why a rainbow is considered universally beautiful, or a butterfly, or certain landscape scenes we see as attractive, but not others.

There is much we simply do not know yet about the human consciousness, or why it does what it does, or perceives things the way it does. So let's concentrate on consciousness, since it may well be the very foundation for reality, like it or not! But let's keep the power of belief, at least on the personal level, in mind as we do. For belief stems directly from the human consciousness.

By the way, the belief in the idea of a greater consciousness than our own permeates every major religion and most minor ones on Earth. The Buddhists, for example, think everything only exists because of consciousness and is a part of that consciousness. So consciousness and belief have always had an extraordinary connection for humanity. And this seems to be the same of late, for quantum researchers, as well.

There is some good news on how we view all this. As one author of an article put it with regard to science, religion, and the study of consciousness in the West versus religion, mysticisms, and consciousness:

"We in the western scientific culture have just begun, en masse, to explore our inner cosmos. Inner exploration has been an intellectual activity in the relatively recent past, has been associated with psychotherapy. Now inner exploration is beginning to enter the domain of emotional and spiritual development as well." -- **Unity of Consciousness Experience, Nature of the Observer and Current Physical Theory, Richard Amoroso**

PART 2—ODD TYPES OF IMMORTALITY

CHAPTER 4—Parallel Worlds or Realities

"I have seen a world within a world within a world but never a world ending without another world."
—Meeran W. Malik

Now we are closing in on the idea of life after death itself, the continuation of our strange consciousness in some fashion despite the death of the human body. One of those ways, a particularly strange way, indeed, is with the Many Worlds Theory, which many quantum physicists think is true, as do quite a number of cosmologists, astrophysicists, and astronomers, as well, if for different reasons.

In fact, there are a number of major theories that demand there be parallel universes or worlds. These aren't just minor theories, but major ones, and again, ones that currently are accepted by most physicists, cosmologists, and/or quantum physicists. So can there be more parallel worlds or universes with different versions of us in them, or even versions where we don't exist at all? Let's see if we can answer that.

Now, without getting bogged down in mathematical formulas or such, let's take a general overlook at these ideas and then see why they might apply to some form of immortality. However, it is important to note, as is stated in the article in **Scientific American (July 19 2011)** titled, *The Case for Parallel Universes, Why the multiverse, crazy as it sounds, is a solid scientific idea, Welcome to the Multiverse*, by Alexander Vilenkin and Max Tegmark, that parallel worlds or alternate universes are not so much theories in themselves, as they are a predicted outcome of other theories. Here are those main theories that involve some possible form of immortality for us:

1. Inflation Theory. The theory the universe inflated from a tiny singularity, "the Big Bang."

2. An Infinite Universe of Many Parallel Worlds. The theory that an infinite universe, one that goes on forever, might include an infinite number of parallel worlds.

3. String Theory and Multiple Universes. String theory implies there has to be more than just three dimension and the most promising form of that theory includes the following, Mbrane Theory.

4. Mbrane or Multiverses. Various universes exist like "branes" of a rather flat, perhaps undulating nature and occasionally they collide causing new universes to be born.

5. Black Hole Universes. A black hole may have a different universe inside of it and even that universe may have black holes in it, as well, so there could be an infinite number of nested universes.

6. Simulation Theory. What you see isn't what you get, but rather is just an illusion of reality, just a simulation being played out in some incredible and perhaps vast quantum computer.

7. Quantum Theory and Multiple Realities. Quantum Theory strongly implies there could be an infinite number of realities where all possibilities have to play out.

8. The Cosmic Crunch. The universe recycles over and over, perhaps eternally, by expanding and then collapsing in a "cosmic crunch."

There are even more theories that include the idea of other universes, parallel/alternate realities, but these are the main ones. And they all imply that there may be multiple universes and even ones with another "you" in them, as well as alternate versions of you.

This is an incredible idea and it does challenge our idea of what it means to be "unique" is, but we can get to that later on. First, with regard to the idea of immortality, these theories are of importance here because they allow for different versions of immortality. I will keep this as simple and short as possible, but again, it is necessary to discuss these theories of the universe to some degree in order to build on them, the idea of immortality, and just how possible such a thing might actually be.

Chapter Conclusion. All these theories of the universe, and these are by far and away the most favored by scientists, lend themselves to the idea that in an odd way, we might just have a strange form of immortality. Some of these types of immortality are strange ones, and not exactly what we would probably prefer, but if we are talking about the "you" continuing after your death in this reality, then these alternatives do imply a form of immortality. They just might not be what you expected….

Let's start with Theory Number 1, the needed concept of Inflation to account for how our present day universe is, in the next chapter. Again, we won't dwell at length on these, but just to give a general idea of each one of them and how they might hold the promise of immorality for us.

CHAPTER 5—Inflation Theory

"After your death, you will be what you were before your birth."
—Arthur Schopenhauer

As most of those who read these types of books probably already know, our universe is expanding and it seems to be expanding ever faster. Like the old comparison to a loaf of rising raisin bread, ever point in our universe is moving away from every other, just like raisins in rising dough would move away from each other (with the exception of clusters of galaxies where gravity is still more powerful than the expanding process and so acts to keep them together…for a while yet, at least).

The universe also appears to be "flat," meaning it has no apparent curvature to it (this also implies it may be infinite with even more important implications). In other words, if you have two parallel lines and send them far out into space, no matter how far they travel, they will always remain parallel and never merge, as they would in a universe with curved space. Also, based on how homogenous our universe is today, how equally (overall) matter and energy seem to be distributed within it, it must have gone through early stage(s) of inflation to achieve this state, and very early on, went through a very rapid expansion phase, indeed. Now, please remember, this isn't just a pie-in-the-sky theory, but is the main theory of cosmologists today.

First achieving notoriety in the 1980's, Inflation Theory, became increasingly accepted because it gave us a means to measure and actually test it. Furthermore, Inflation Theory also allowed us to make predictions based upon it. Since then, various observations of the universe, along with predictions made of what should have resulted if the theory is valid, have borne Inflation Theory out. In other words, the theory seems to be correct, based on data, various forms of evidence, testable predictions, and even observed evidence by astronomers and cosmologists.

However, and this is important; Inflation Theory of the universe also requires something else, something less testable at present, but it has to be there in order for the theory to work and that is, multiple universes! Along with Inflation Theory's necessary partner, Quantum Theory, (often referred to as Quantum Mechanics), there must be multiple "worlds," or universes.

Since Inflation theory has been tested, has allowed us to make predictions based upon it, and these are all holding up well, we have to assume that Inflation Theory is probably correct. It predicts quite well the universe we see today and its many attributes and properties. Therefore, even though we can't readily test for there being other universes yet, as well, we think we "know" they must exist based on this. Simple logic states that:

1. Inflation Theory won't work without multiple universes being involved.
2. Evidence strongly indicates that Inflation Theory is correct.
3. Therefore, multiple universes must exist!
4. If multiple universes exist, then there may well be a form of immortality as a result.

It's as simple as that. If you have enough universes similar to ours, some will end up being the same or just ever so slightly different. Therefore, you would exist in more than one universe and other versions of you might just exist, as well. In fact, according to the laws of probability, this must be so, that is if other universes are also the result Inflation Theory.

So there is a form of odd immortality in that even though you might die in this universe, "you" could live on in others along with even other versions of you doing the same thing. It would appear you just might be a hard person to kill, given the number of universes in which you may exist, and so this would constitute a form of immortality. Die in one, continue to live on in another.

And since not all universes may be at the same exact age level, some with a slower rate of developing than others, there may even be universes in which you haven't been born yet, but may in years to come. Again, a form of odd immortality. Perhaps not the most likeable one, since we would prefer to continue on with what we know as of now, an uninterrupted existence, but even so, a form of immortality nonetheless, and this is no less in comparison to the idea of reincarnation, where you don't remember your past lives either.

And remember, Inflation is the accepted theory by the vast majority of scientists and researchers right now. This means there is a distinct possibility, even probability that there might be more than one universe with you in it! So having dealt with that, let's move on to the next theory, that our universe is infinite. This, too, allows for a strange form of immortality.

CHAPTER 6—An Infinite Universe

"Surely God would not have created such a being as man, with an ability to grasp the infinite, to exist only for a day! No, no, man was made for immortality."

—**Abraham Lincoln**

Remember how we mentioned that our universe appears "flat" without any curvature (overall that is, except as in distorted space-time locally caused around suns, planets, black holes, etc., because of their mass)? To reiterate, this means that if you sent two parallel lines out into space (as mentioned above), they would never meet. Also, scientists are convinced our universe is far larger, in any case, than we can see. Since the beginning of the universe, light has been traveling for about 13.7 billion years. So we can "see" (via telescopes, etc.) only that far in any direction.

However, because the universe has been expanding due to inflation, and is still inflating at an ever faster rate, many scientists feel the universe has to be at least 93 billion light-years across, at the very least, although most of that we can't see, of course, because light hasn't had time to reach us yet from such a vast distance. Nor will it ever, probably. At the rate the universe is expanding, those points farthest away from us will expand away faster than light can travel, and so we will never see them, because the speed of light will never allow their images to reach us.

In fact, it is estimated that around 20,000 stars every second disappear forever on the other side of that speed barrier, and we will never be able to see them again. It is rather like having a car racing toward you on a highway at a certain speed, say 60-miles per hour. But the road is a magic one and it is stretching out at a faster rate than the car is traveling, say, at 100 miles an hour, but every second, even that speed is increasing. Try as the car might, traveling at just 60 miles an hour toward you, it will never reach you because the road between you and the car is stretching out faster than the car is traveling, so the car is actually getting farther away from you with every passing second until eventually, it disappears over the far horizon, never to be seen by you again.

Another way to view it is, as with those horror movie scenes, usually dream sequences or such in them, where someone starts running down a short corridor, but even as they do, the corridor suddenly keeps stretching farther and farther out ahead of them and they can never reach the end because of this. Think of the original Poltergeist movie with the mother trying to reach her child's bedroom down the hallway, for instance. That's how the universe is behaving because of its expansion.

Moreover, many scientists believe the universe isn't just that 93 billion miles or so across, but is infinite, and will expand infinitely. If space is flat, as it seems to be, then this would be so.

Why is this important? Well, oddly, because of subatomic particles. Subatomic particles make up everything that is made of matter—everything! Like Lego blocks, they can be assembled in an incredible number of ways, an absolutely huge number of ways. However, the number of ways the particles can be assembled does have an ultimate limit, although a very high one. There is still only a finite number of ways one can assemble those particles before repetition sets in.

This is because there are only so many types of particles and there are only so many ways they can come together. Again, that number is NOT infinite. This means that eventually, they have to repeat what they assemble, in various ways, alternate versions, and sometimes in exactly the same ways. Given an infinite universe and vast amounts of time this would, according to the laws of probability, not just be possibly true, but actually has to be true! Those particles would assemble in every possible way, every variation conceivably allowed by the laws of physics.

Think of those Lego blocks again. You can build cars, houses, bridges, towers, planes, whatever you like out of them, but if you only have so many such blocks, eventually, given enough time, you have to start repeating yourself in what you build. The same holds true for the universe.

So if the universe is infinite, things will repeat. Go far enough "out there," and there should be all sorts of variations on our own Earth and solar system. There will even be some exactly like ours, others slightly different, and there will even be ones with a "you" and different versions of "you" on those other Earths. Theoretically, an infinite number, where you are exactly the same living the same life, and/or other versions of you living alternate lives. In some you will be rich, others poor, some with certain family members, some where there might be a sister or brother who has never have been born, etc. There would also be some in which you haven't yet been born but will be, and others where you've already died. Every conceivable possible variation would have to play out. Endlessly!

Therefore, in essence, we have other universes or realities right within our own universe; perhaps incredibly far apart, but still just as real as our own and in this same universe. Again, the logic is straightforward.

1. If space is infinite (and it seems to be based on what we know so far), then things must repeat since although the universe might be infinite, the numbers of ways things can be put together with subatomic particles is finite so again, they must eventually repeat. And there would be all sorts of variations on that theme, as well.

2. If things repeat infinitely, sooner or later there must be an exact repetition of our sun, earth, other planets of our solar system, etc., including even ones with another version of "you" on that Earth. Or perhaps in some cases, exactly the same as "you," as you are here, with no discernible difference.

3. If this is so, then that means there are, indeed, a version of parallel worlds, similar or even the same as our own "out there," but still in our own universe. Some would have "you" living on that Earth, others where "you" might never have been born but everything else is the same, and still others where "you" live different versions of your life—again, some where you are rich, some where you are poor, some where you are ill, some where you are healthy, etc.

An infinite number of parallel realities within our own infinite universe would seem to have to be true. All probabilities, all variations, would have to exist and play out, including all possible variations of "you." So again, we have an odd form of immortality where we might exist in the here and now, but "elsewhere," as well, and not only in the "now," but in the "past," and in the "future, since not all possibilities would have to occur simultaneously. This creates a de facto form of immorality for the you that is you!

Chapter Conclusion. One thing is certain; if space is infinite, that is, the universe goes on forever, then most assuredly, you exist in more than one "place" and probably in different times, as well. The current "you" popped into existence almost 14 billion after the universe began here. Others may have "propped" in, in some far distance region of the universe at the same time, as well. Still other versions of "you" will pop into existence at a later date. Not only that, but perhaps an infinite number of versions of you, leading slightly different to very different lives from you would also exist and at different times.

All that is needed for this to happen is an infinite universe and the laws of probability to hold true throughout it all and it's a done deal. Since many scientists and researchers think the universe is infinite and subject to the same basic laws throughout, then all this would be true and real! So how does that make you feel? How does it change your view of things to know that you exist in multiple places at the same time and at different times, in the past and in the future? You may not be consciously aware of these other versions of you, but they would exist. So this creates an odd form of immortality, since there is more than one version of you, many "same versions" of you, as well, and some yet to be born, and some that have already lived and died. An incredible idea that could well be most likely if current theories hold to be true. Now let's turn our attention to yet another theory of the universe that also requires more than one "you" in it.

CHAPTER 7—String Theory

"When we understand string theory, we will know how the universe began. It won't have much effect on how we live, but it is important to understand where we come from and what we can expect to find as we explore."

—Stephen Hawking

String Theory. This theory is a mathematically complex one, but the idea in its basic form is simple. It states that all subatomic particles that act to make up all matter are actually made up of "strings" of energy, or loops of such energy (depending on which scientist one talks to).

So besides the fact that this means we are all really just a sort of a form of "condensed" energy, it also means something else. String Theory predicts that there has to be entire other universes. That number is approximately 10^{500} different types of space-time universes. This is a fantastically large number of universes, and some theorists say this is just the minimum required for the theory to work, that perhaps there are even many more than that. Just try multiplying that out some time! The number of zeros in that number will stagger you!

Moreover, each of these universes would, or could have their own laws of physics, but not necessarily all. Some could, theoretically, repeat, so there may be more than one of "our universe," as well, or possibly many variations on it. Therefore, it is theoretically possible there could be more than one "you" or "me" existing "out there," or perhaps even many versions of us. Scientists simply don't know for sure.

What they do know is that one of the most accepted theories of the universe seems to require multiple universes in order to work. It has to have them. Just how many, seems to be the question but even at the minimum, it's still a staggeringly high number of universes. Again, we have the idea that other universes exist, many completely different from our own, but again, some similar, or perhaps even alternate versions of our own. Statistically, it would be probable. Moreover, those could well have "us" as we are now in them, or again, possible variations of each of us. So once more, we have the idea of a form of odd immortality, where we exist in the here and now, but also "out there" in other universes.

Chapter Conclusion. One thing to note with the String Theory; if all particles of matter are really just loops of energy vibrating at different frequencies as the theory claims, then that means we humans are really just a collection of vibrating energies, merely manifesting as a solid being. So could one define our true state of being as that of pure, vibrating energy? If the String Theory is correct, then yes, that is what we ultimately would be.

So perhaps the idea of some psychic mediums might not be far off. Perhaps different "planes of existence" are just different levels of vibrations as they sometimes put it. After all, if String Theory is correct, that is what the physical universe could be defined as! In any case, it's an intriguing supposition and seems to follow logically from the initial premise that all matter is made up of vibrating strings of energy.

Now for the next theory, Mbrane Theory and what it means for all of us. By the way, String Theory is very popular among scientists, but there are many versions of it, and so those scientists have been trying to devise a theory that would bring all the different versions together. They think they might have found it in Mbrane Theory, although this theory as yet, is still incomplete.

CHAPTER 8—Mbrane Theory

"In some parallel universe, there exists another you and also this message. That version of you is thinking about you right about now."
—Rajesh

Mbrane Theory of the Universe. This theory is an attempt to bring all the different versions of String Theory together. It posits many things, but one of the things the theory suggests is that there a multitude of universes, again, "out there." They have their own space-time and existence. They float along in a sort of non-space environment (a generality, but it suffices here), and occasionally, they bump into each other. When they do, at the point of contact of such collisions, new universes are birthed, or new "branes," as they are called. This can go on forever and probably has if the theory is accurate, meaning, again, there must be an infinite number of such universes out there with new ones being created all the time.

Furthermore, it probably means that some of those universes by sheer chance, could be other versions of our own. Some scientists say that since they would be birthed from the same parent universe or universes when colliding, they could well be very similar, depending on which researchers one talks to, parallel universes may be worse than common.

What cosmologists like about this theory is that it neatly explains what came "before" the big bang that created us. Our universe then, is just the result of a collision between two other universes, or "branes," again, as scientists refer to them, and that those two already existed. Our universe would be an offspring of them, so to speak.

Again, if an infinite number of such universes exist, that means all probabilities could play out sooner or later, including other universes that are so like ours that "you" might exist in them, as well. So again, there might be more than one "you" and even many versions of you scattered throughout the multiverse. Moreover, since universes may have been forming forever this way, then you might have existed in some long past universe, as well as this one now, and again in a future universe yet to be created. There may be no end to the number of "you" that might exist in the past, now, and forever in the future.

Chapter Conclusion. I am keeping some of these chapters on the various theories about our universe necessarily short, because my purpose here isn't to give my reader a crash course in cosmology, but rather to illustrate how, in many cosmological theories, they allow for there being more than one "you" in existence, thus allowing for a form of immortality since you theoretically may have always been alive somewhere "out there," as well as here now, and in universes yet to be born.

One thing; it is intriguing to see just how many theories may allow for this odd type of immortality and if even just one of those theories is correct, then the answer is there are multiple versions of you, again stretching into the past, here in the present, and somewhere "out there" again, in the future.

This sounds almost godlike in a way, doesn't it? After all, in the Catholic religion for instance, part of a prayer says, *"one God, now and forever…world without end. Amen."* We may not have the power of God or gods, but we do seem to have the longevity of such if any of these theories of our universe/multiverse are correct. We, as individuals, just may be *"now and forever,"* as well. Nor does it stop there. There is another theory that is similar to this one called the Black Hole Theory of universes. This is an intriguing one, also, and again offers us yet another chance at an odd form of immortality.

CHAPTER 9—Black Hole Theory of Universe Creation

"It is time we Postulate that a Black Hole is Some kind of Womb."

—Donnie Harold Harris

Black Hole Theory of Universes. This theory is intriguing and it offers yet another way for immortality of an odd sort. Scientists have noticed that our universe could well have the same exact description as that of a black hole. Our universe could well be bounded by an event horizon, so we cannot escape it, and just like a black hole has one. Our universe was created, supposedly, by an initial explosion of energy known as the Big Bang, just as black holes are created by the same sort of thing. When a star or enough matter collapses in on itself, there is an instant where it reaches a point where a black hole is created be a big implosion.

We can't view what is beyond the event horizon of a black hole because nothing can escape one, but we know that an incredible amount of mass has collapsed to a mere point in space-time in our universe to create such a thing. Even so, again although beyond our limits to comprehend, that mass has to be somewhere still even though probably converted to pure energy! Did all that energy then create an explosion and its own space-time continuum? After all, again, that could be a description of Big Bang that created our universe—a sudden eruption of energy from a singularity that went on to expand and become our universe. We just don't know the answer to that. Standard physics breaks down and cannot account for anything within the confines of the singularity that lies inside the black hole's event horizon. But we can make educated guesses.

Some scientists theorize that black holes are new universes, and that within the inner confines of them, a new space-time is expanding and that this is how new universes are born. In fact, our universe has given rise to an incredible number of such black holes, and these same researchers conjecture that even our universe itself was created from a black hole in exactly the same way. Black holes forming within black holes, as it were.

Some scientists even go so far as to say that universes that form black holes within them are more likely to far outnumber those that don't. It is almost a sort of evolutionary process. If a universe is very different from ours, it probably won't produce black holes. Eventually, that universe will die by one means or another without having then produced any "offspring" universes.

However, if similar enough to ours in nature, a universe will then form and go on to produce black holes, even as ours has done. Eventually, as a result of this ongoing process, those universes like ours that create black holes would go on creating many more universes in the same way, until they far outnumbered universes that didn't produce black holes. Moreover, this whole process may have been going on forever with one universe spawning numerous new ones, and then those spawning more, etc., in this same way, ad infinitum, ad aeternum.

If these universes have to be very similar in nature to each other in order to do this, create more black holes, it could well be that some that are born that are even identical to ours and so might have another version of Earth and "you" in them. If the number of universes is truly infinite, and this has been going on forever, then this would be a real probability and not just a remote possibility.

Sooner or later, variations on a theme should occur, even as repetition would, that is, if these new baby universes have to be very similar to ours in their laws of physics in order to produce black holes themselves. And where there is infinity, there probably has to be more than one you.

Chapter Conclusion. The Black Hole Theory of the creation of universes by parent universes is considered a very real possibility by many scientists, simply because the definition of a black hole also so well describes our own universe, too. So the idea of worlds within worlds, each expanding to infinity by creating their own space-time as they go, is an incredible one.

In that infinity of endless creation, there would be some universes formed in the past (compared to when our particular universe was created) where you could well have once existed, as well as in the present, and some universes where you would again exist in the future. There may well be universes existing simultaneously with a "you" in them, as well. Surely, a form of immortality with being able to exist always and with multiple versions of you perhaps existing at the same time.

Again, with all the metaphysical implications this concept entails, I can't help but be reminded of that one part of a prayer, *"...world without end. Amen."* Amen, indeed. The saying that usually ends most prayers. "Amen," when translated from the Greek Old Testament version, most closely means "so be it" in English. Amen has also been translated to mean "surely," or "verily," as well. Is it not rather satisfying to think that we can exist not just from now and into the future in various universes, but that we may have always existed somewhere and some when, even if we aren't "consciously" aware of it. If this theory of the universe being form by a black hole is correct, then all this could well be so. And if so, then so be it. The possible answers science seems to be giving us late to tend to have a rather metaphysical echo to them of late.

CHAPTER 10—Simulation Theory

"The strongest argument for us probably being in a simulation is the following: 40 years ago we had Pong, like two rectangles and a dot. That was what games were. Now 40 years later we have photorealistic 3D simulations with millions of people playing simultaneously, and it's getting better every year."

—Elon Musk

Simulation Theory—The "Matrix." I'm sure we are all well aware of this theory of the universe, that it is a simulation created either by our descendants at some future time to recreate their own past, or perhaps by some future extraterrestrial species that might want to better understand how (by their time) extinct humans may have existed and what might have happened to end that existence.

For me, there are other theories, as well. With the advent of computer games, once being so simple as a single pixel bouncing bounced back and forth between two lines of pixels (Pong), to what they've become now, it could well be we are living in an elaborate video game. We could either be NPCs (non-playing characters just added as background, but conscious ones so as to seem more "real"), or we could be people in the future or some other reality that have entered the game to play along in it but with the proviso that our other and "real" existence cannot be remembered for the duration of our stay in the simulation. This last would make for a total immersion effect, where the people playing as characters in the "game" would think they actually were the characters.

We could have been made to forget our "real" lives "out there," outside this simulated universe while we are playing. Think of either version of the movie, *Total Recall*, where people just have memories of a vacation implanted in them, rather than going on an actual one, or perhaps like the movie, *The 13th Floor,* where they created simulations of past cultures (1930's – 40's Los Angeles, for example), just in order to experience and enjoy such simulations. The characters in those simulations were conscious, but not aware they were only a simulation. And just maybe, in order to experience it all to its fullest, again, that is why we are made to forget our other and "real" existence outside of this one.

So if this is just a simulation, what does that mean for us as individuals stuck in it? Well, for some, such as the billionaire, Elon Musk, it means striving to find a way or means to get out of the simulation. Supposedly, he has even hired researchers to help achieve this goal. Although personally, if I were a billionaire, I think I might just want to stay in this simulation....

Even so, it seems a lot of people hate the idea this might be a simulation. Therefore, we should give some thought to how likely it is that this theory might be true. Nick Bostrom is credited for having come up with the idea of Simulation Theory in its present form, but others have been involved, as well. However, being the foremost proponent of the idea, let's concentrate on how he sees it. He states in a very matter-of-fact and logical way that any of the following might hold true (according to **Wikipedia**):

"One thing that later generations might do with their super-powerful computers is run detailed simulations of their forebears or of people like their forebears. Because their computers would be so powerful, they could run a great many such simulations. Suppose that these simulated people are conscious (as they would be if the simulations were sufficiently fine-grained and if a certain quite widely accepted position in the philosophy of mind is correct). Then it could be the case that the vast majority of minds like ours do not belong to the original race but rather to people simulated by the advanced descendants of an original race."

It gets worse. If some of those simulations have been running long enough, and the people "in" them are conscious and can learn, as we seem to do," then sooner or later, they, themselves, will create their own simulations with conscious beings in them, too!

Worse yet, those simulations might then go to create more simulations, which in turn then go on to create even more simulations *ad infinitum*. These would be called "nested simulations," being simulations within simulations.

As difficult as that might be to comprehend, it could well be true.

"There's a billion to one chance we're living in base reality."

This is a quote by Elon Musk. Other researchers think the odds of us living in a simulation are around 33% to 34%. Still others say the odds are even more likely this is a simulation. Moreover, some of them seem to think these probability assessments will eventually go even higher than they are now as we learn more about our universe.

One of the things about the simulation theory is that no simulation could be as perfect as "natural" reality, since no computer, regardless of what reality it may be in, could recreate all of its own reality perfectly. There would be certain problems that would manifest in such a simulation:

1. One problem would be that just as with images in computer games, our reality would tend to pixelate on the smallest scale of things. Just as if you were to magnify the image of a computer game monitor enough in size, the visual of the game would then turn into separate pixels of light. It seems that so, too, would our reality. In other words, on the very small scale, our reality should pixelate. And again, it seems to. When we go down to the super small level of our reality, it, too, pixelates. Matter becomes molecules, which in turn are made of atoms, which in turn are made of subatomic particles, those smallest bits of matter that make up everything, What's more, according to String Theory, even those particles might be nothing more than energy loops, even as the image on a TV screen is just the result of photons—energy.

Time also seems to pixelate. There is a theory that the smallest unit of time is a "jiffy." Or as one source (see References) says:

"The jiffy is the amount of time light takes to travel one fermi (about the size of a nucleon) in a vacuum. ... Theoretically, this is the smallest time measurement that will ever be possible. Smaller time units have no use in physics as we understand it today."

Nor does the pixilation of our universe stop there. There is even a smallest length of measurement possible, and nothing is supposed to be possible that is smaller than that. It is known as the Planck limit. As **Wikipedia** defines it, a Planck length is:

"...is equal to $1.616255(18) \times 10^{-35}$ m. It is a base unit in the system of Planck units, developed by physicist Max Planck. The Planck length can be defined from three fundamental physical constants: the speed of light in a vacuum, the Planck constant, and the gravitational constant."

Now, that is incredibly small! So when it comes to Simulation Theory and the idea things in our universe would have to show evidence of such pixilation to back the theory, we have that. We have it regard to the smallest unite or "piece" of time. We have it with regard to the smallest bits of matter, subatomic particles, and we have it with the smallest length possible, the Planck Length.

2. Glitches. There is something else we should have, too. Just as even the most elaborate video games do, our universe, if a simulation, should have occasional glitches. Remember the movie, *"The Matrix?"* It had a cat that walked by a doorway twice in rapid succession, a "glitch." Moreover, we do seem to have such glitches, as well.

If you doubt this, just check out some of the videos on YouTube regarding this and see for yourself. Yes, as always, some can be faked, but others are not. From jets seemingly frozen in the air, to seeing two identical people on a subway car, to people driving in a car and seeing everyone outside seemingly frozen in place, there are lots of such videos of glitches. Some people will see the exact same car, license plate and all, following its doppelganger on the freeway and even have filmed it. Others watch as cars ahead of them just fade out. This has happened historically with people, as well, where multiple witnesses have looked on as a person simply disappeared from view.

Déjà vu might well be a mental glitch in the universe. Who knows? Perhaps the character you are right now is having a bit of a memory from having been a prior character at some other point or time in the simulation?

Yes, this last seems highly suspect as proof of a simulation, but some of the evidence seems almost irrefutable. Furthermore, the pixilation part of it all is certainly real enough, as far as physicists and quantum physicists can determine. I do suggest you watch some of those videos, though of such glitches. Some obviously have mundane explanations for them, but others do not seem to. Judge for yourself.

Now, what does all this have to do with human consciousness and life after death? Well, first, if this is a simulation and you are in it as a nonplaying character then there must be a program, some subroutine that creates you that would exist even if you died in the "game." Furthermore, if you're a gamer in the game and had your memory of any other existence temporarily wiped so as to make the game more real to you, then you would also exist "outside" of the simulation, as well. Finally, if you are a consciousness in the game, you can be, and might well be, recreated at different times by the "programmers" for different simulations besides this one. This is another odd form of immortality, where again, one does not realize it, but nevertheless, "you" might go on in one way or another.

100% Probability This Is a Simulation? Fouad Khan on April 1, 2021, in the respected journal, Scientific American makes the following case for the chances of us being in a simulation are 100%. He bases this conclusion on the following logic:

1. If we are in a simulation, there would be all those pixilation of things on the small scale, whether time, distance, or subatomic particles. We have that.
2. Glitches would and could appear, and there seems some evidence for this phenomenon happening, as well.
3. If we are in a simulation, it would be very hard to prove, because everything we relied on as proof is also a part of this simulation, with the exception of one thing that would remain constant, unchanged,

and immutable. That is the processing speed of whatever device the simulation is being run on. That cannot be altered in the simulation and so would form a bedrock and unchangeable constant. We would seem to have that, too. There is one such immutable constant as to how fast things can process or travel, and that is the speed of light. At approximately 186,000 miles per second, the speed of light is a constant. Is this, then, the processing speed at which our simulation is running? Scientists have no idea why the speed of light is this particular speed—it does, in fact, seem arbitrary, but even so, it exists and is a constant.

4. Since we have the required constant of a processing speed that can't be disguised or changed in the simulation, in this case, the speed of light, then there is, according to Mr. Kahn, a 100% probability this is, indeed, a simulation.

Chapter Conclusion. Does this Simulation Theory invalidate you as a person, as a consciousness, since you might just be a programmed character in a simulation or game? No, it does not. Consciousness is unto itself. Whether a simulation, reality, or whatever, you are conscious. You are aware of your own existence. So in that sense, you are real enough, at least. You are conscious of your existence.

In fact, one could argue that being aware of yourself, you are just as conscious as any "programmer" or "gamer" who had created the simulation or has participated in the "game."

You are no less aware than they would be, and therefore, would be on equal terms with them in that one regard. Even if the game or simulation were terminated, the subroutines, the programming that created you would still exist in all likelihood in some type of computer memory somewhere. What's more, "you" could be then reactivated at any time, or if there are multiple simulations, you might be "running" in more than one of them at any given moment. Again, an odd form of immortality.

In any case, we have now covered a variety of theories about the universe, mainstream ones that many scientists adhere to as being the probable actual state of things for "out there," as well as "down here" on Earth. All of them point to the idea that there could be multiple, perhaps even an infinite number of "you" existing and not only simultaneously, but in the past and in the future of different universes. There could be millions, billion, trillions, or even more of "you."

Now, besides raising the question of whether this takes away one's uniqueness, it does raise the distinct probability that even if one or more of "you" die, there are many other versions of "you" that will continue to exist. In many case, this could be the same "you" exactly, as with the idea of the universe being infinite and the Laws of Probability having to play out sooner or later in all their diverse outcomes.

Think on this a bit. If you die "there," you are still alive "here," or vice versa. You may not realize it consciously, but if not on the conscious level, then perhaps on some other level you might be at least partially aware of this fact. What is déjà vu really, for instance. Scientists have come up with one theory but no real proof to back it that it's a delayed (synaptic?) response that tricks your brain into thinking you have already experienced the exact situation before.

However, this is just pure conjecture on their part when they say "this might be the cause of déjà vu." They have no empirical data to support the conjecture at all! But if a scientist says something is so, that in itself often seems to make such a mere conjecture a fact...or not.

Also, if your existence were to terminate here, there is another universe or reality, parallel world, or whatever where you go right on existing, so for that person, you never ceased to be at all. And since those other "you" out there would, in essence, be a continuation of you, just in a different location or space-time you would have an actual form of immortality even if you didn't consciously realize it.

In other words, one way or the other, you might get to live out your life and it might be a very long life, indeed, perhaps endless, because some realities (not all having been created at the same time) would contain another you long after you should have died of old age in this one.

Yes, it sounds improbable to the rational mind, but think about it. If any one of these theories is valid, let alone most or all of them, then this must be so! The Laws of Probability say it must be so. Flip a coin thousands of times and it will come up either heads or tails because it has to according to those same laws.

But once in a while, if very rarely, the coin will land on its edge. In just the same way, according to those same Laws of Probability again, you must be recreated or exist elsewhere. This, then, forms a sort of immortality. Even though dying here, you still exist "there." Even if you aren't directly aware of this fact.

PART 3--THE QUANTUM SOUL

CHAPTER 11—A Quantum Consciousness Instead

"Universe is like a soup of consciousness. Brain is just a receiver. Awareness is continuously transforming thoughts into matter."

—**Amit Ray,**
Enlightenment Step by Step

We have explored the idea of consciousness and what it might be. We have also explored how different versions of "you," and even the same "you" might exist in parallel worlds and through a variety of means, all involving major theories of the formation and function of the universe, many of which are mostly accepted by various cosmologists, physicists, and quantum physicists, alike. Admittedly, some of these theories have more adherents than others but the truly so-called mainstream ones have many. They are a majority of scientists, in fact.

All of these mainstream theories imply that there is more than just our universe/reality, and that you almost have to, according to the laws of probability, exist just as you do now, or in different variations of your current life, within some of those different universes and/or realities. So in a very real sense, you have a sort of immortality if not one you are personally aware of.

However, many of us don't want to just have a different version of ourselves continuing, even if it is the same one as us, and not be aware of it. After all, we don't know anything about "them" or that they even exist, and if so, where and when, so it doesn't feel as if they and us are the same being. Many want a continuity to their lives as immortal beings.

So what about us? What about our own unique consciousness? Even if another "me" is having all the same thoughts at the same time as I am but somewhere else, I'm not personally and directly aware of this, so again, as far as I'm concerned, there is only the one "me" for all practical purposes.

Therefore, if I die and when my brain stops working, is that the end of me in this reality at least? Well, the answer, for a number of reasons, is "maybe not." This may not be the end of you and in fact, more precisely, this answer should be rephrased as most "probably not." We will go into a number of reasons why this might be true, but one of the main ones is the idea that our consciousness may be quantum in nature.

Now, don't let this term scare you. We're not going deeply into quantum physics at all, but instead we're just going to show how it applies to immortality in a step-by-step fashion, and hopefully, clearly laid out in this regard.

Does human consciousness have a basis in the quantum realm? For a growing number of researchers who often refer to themselves as quantum biologists, or at least to their research as quantum biology, it appears our consciousness might be based on the weird world of quantum physics. Remember how I touched on this idea when I spoke about microtubules in the brain that seemed to work in a quantum way? Well, there is more than even that involved.

1. Many scientists think that the human mind cannot be fully explained by the physical anatomy of the human brain. That is, the physical neural network in the brain can't seem to give rise to the mind, it's choice of thoughts, even thoughts themselves, or the concept of choice as in free will or randomness of thoughts. Therefore, those researchers proclaim that although the human brain is solely subject to classical physics and not quantum physics (being part of the macro world, and not the micro world of subatomic particles), the brain can only sort of simulate behaviors that would seemingly have to arise from quantum physics.

2. Quantum biologists and various other researchers think the idea above (in Number 1 of this Chapter) is simply not the way to go. Their attitude is that if it "looks like a duck, walks like a duck, and quacks like a duck" it probably is a duck." They also point out that the Principle of Occam's Razor says that if there are two explanations that satisfy a problem the simplest is usually the correct one. In other words, why bother have classical physics "act like" or simulate quantum physics to explain the human mind, when simply using quantum physics is simpler and more direct?

3. Besides the argument set forth by quantum biologists in Number 2, those same researchers say that having classical physics trying to explain the human mind by saying that classic physics can sort of simulate the products of quantum physics isn't good enough. They argue there is just no way that classical physics can simulate what the mind does well enough, so it must be quantum physics that explains the human mind. Therefore, human consciousness has to be based on quantum physics in order to do all that it does. The human mind is a quantum state or at least the result of one, according to them.

So just what are we saying that quantum physics can do to create the human mind? Well, on the quantum level, subatomic particles can exist in more states than one, being in two distinct states simultaneously. Quantum biologists say this happens in the brain, in those microtubules. They go in-depth on the molecular nature of those microtubules, the proteins involved, overlapping probability clouds of particles, etc., but we don't need to go into any of that here. Suffice it to say that the human brain does seem to have a section (those microtubules, and perhaps elsewhere) that allows for a superimposed state, thus being in more than one state at a time.

Secondly, quantum physicists say that on the quantum level no particle's exact whereabouts can be found without a measurement being taken, that a particle exists as a wave, or a set of probabilities as to where it might be at any given moment. It isn't until an "observer" checks by direct observation as in a measurement, that the particle collapses from the wave state and its location can then be measured.

If this is true of the human brain, this means that thoughts might be the result of quantum states, having multiple thoughts and the "I" or "me" (the Id?) decides on which thought to think and/or follow through on. It also means our minds may be, to a degree, nonlocal, that is in no one set place.

A bit complex, I know, but one might think of the mind as the "wave-like" state of the physical brain, just as a particle can also have a wave-like state. Whereas, the physical brain is the particle state. If this is so, then the mind may be a quantum thing and be nonlocal in nature, meaning not measurable as being in any exact location.

We can go further. Many researchers now claim that the human consciousness cannot be just a result of the physical human brain, that it can't be a purely cause and effect situation, with the brain's activity producing the mind. Why? Well:

1. How humans and animals can "smell" odors would seem to be a function of quantum behavior, not standard classical physics behavior.

2. Sight does not seem to be just a function of the physical human brain. How do we know this? Because scientists have completely mapped the entire sight system of humans, from where light enters the eye, hits the retina, travels up the optic nerve and into the brain. What they found is surprising. There is no place in the human brain where what we "see" all comes together! There is no brain equivalent of a projection room, or theater for viewing what has been transmitted to the brain. For example:

When we see an object, we see both its color and shape. Let's take a red ball. The eyes send this information into the brain of the red ball, the color and shape of the object, the visual representation of a red ball as it were. This takes a fraction of a second for the signals to travel the optic nerve, but nevertheless, there is a time delay in what we are seeing. Even stranger, the area of the brain (and scientists now know this for sure) that stores the shape of objects we see is in one area of the brain, near the middle to the back of the brain. The area of the brain that stores the color of what we see is much closer to the front of the brain.

HOWEVER, there is no area in the brain (and researchers also are very sure of this fact now, as well) where the two things come together in once common physical area, no region where the shape (a sphere) and the color (red) have a shared physical region where they join up. Instead, the information is stored physically in two widely separated areas of the brain, never to meet. Again, and this is important: **There is no physical location of the brain where the two visual aspects of an object, its color and shape, are united as one.** This is a simple fact.

SO HOW DO WE SEE A RED BALL? We might see a round shape, or the color red, but how is it we see the object in its totality, as a "red ball?" If the mind is an insubstantial result of the physical brain with no physical powers or capabilities to act on physical matter, then how can it manage to unite the image of the sphere from one part of your brain with the color red from another, even though there is no actual place in the brain where this is physically accomplished?

Yet, the mind does this, and it has been proven this isn't as a result of the physical brain doing this itself, apparently. Somehow, something else occurs. That is, the mind, a purely nonphysical consciousness. unites the two things "outside" of the parameters of the physical brain, so that we "see" an image of a red ball as one united thing. We can then store this as a whole memory, but only after the mind creates the image in the first place for us to do so. And as we've seen with optical illusions, sometimes our minds get the wrong image, one that is not a direct, actual representation of what is really "out there."

How this is done, when the process would seem to have nothing to do with the physical nature of the brain itself, is unknown, other than the consciousness seems to somehow manage it in a purely nonphysical way. Apparently, then, this must happen in some quantum way, a way not subject to classical physics, because classical physics simply can't explain it.

Something else to consider: research has demonstrated beyond a doubt that the human mind (not the brain), has a direct effect on the brain itself. Give a human a placebo (a sugar pill for instance) instead of a real medicine and the consciousness, the mind, as it were, often perceives it as a "real medicine," even though it's just a sugar pill.

This then results in the manifestation of positive physical changes in the human body. The person can often get better because the simply believe the sugar pill is real medicine, not having been told otherwise. The reverse can be true, as well, if someone strongly believes in something like Voodoo for example, that could end by causing a negative impact on a person's health. People have believed so strong in curses and such, that they actually fall physically ill and some have even died! The placebo effect goes further; it makes for physical changes in the brain, as well as the body!

In fact, when a human sees something, the amount of data going up into the brain, doesn't compare with the amount of feedback that is given! How can something as insubstantial as the human mind, just supposedly a collection of non-physical thoughts, change something as physically solid and real as the human brain or the human body? It should not be able to do so, if it is just a natural function of the brain but with no physicality. It would be as if one could pick up a physical object and manipulate and change it without any physical means of doing so. Some claim that the human mind is even capable of doing that!

The easiest way to understand this is to fall back on the earlier mentioned idea of comparing the physical human brain to a movie projector. Picture the human mind then, as a movie projected onto a screen from that physical projector. Your mind is just an image playing out on a screen, so to speak. According to some researchers, the mind is not "real," in that it has no physical substance and so shouldn't be able to act in any physical way on its surroundings, just as the movie on the screen can't act on its surroundings because it is just a projected image, composed of light photons and nothing more.

The projector (the physical brain) generates the image (the mind), but the image can't alter anything about the projector. It is merely a product of it. And if the mind is just a projection of the physical brain, how then can it make changes in the physical brain? It shouldn't be able to, any more than a movie in a theater could change the shape or structure of the theater or projector that is projected it.

If the mind is strictly a function of the physical brain in this way, as the classical physics approach implies, and so is just a consequence of the brain, then it simply should not be able to affect the real physical brain projecting it in any way whatsoever. It should be a one-way street only, projector sending image to screen, as it were, but the screen unable to interact with the projector in reverse. But the human mind is not, apparently such a projection. It does somehow affect the brain often and to great extent, as well as the human body, as well. In fact, it's effect on the human brain and body appears to be an ongoing and constant thing.

The human mind is capable of speeding up or slowing down the heart rate or breathing rate for example. It can speed healing processes in some cases. The mind can create mental states that result in physical states, such as paralysis, catatonia, or just the opposite, giving the body the ability to perform incredible feats, as with those who meditate being able to achieve incredible things with their bodies at times, such as slowing their breathing way down, or being able to withstand extremely cold temperatures with no resulting frostbite or adverse physical effects. There is a man who can run miles, barefoot, and close to being naked in subzero temperature for hours and have no negative repercussion—no frostbite or frozen fingers or toes, for example.

Remember those stigmata? They must be created by the mind, because it certainly isn't the physical brain doing that, at least, not initially. The mind is quite capable of rewiring the neural network in the brain although based on the classical physical interpretation of the mind this should be impossible.

Moreover, this ability has been shown repeatedly by researchers when people have strokes. Therapeutic exercises are based on this fact, where the stroke victims concentrate on changing how their limp arm or leg works and how well, and with long practice and a lot of hard work, it results in actually causing rewiring in the neural network of the brain. This often has helped those with stroke-injured, crippled limbs to function better. This has been proven repeatedly! The mind can rewire the neural network of the brain to compensation for problems and damage.

But if the person with the stroke doesn't do this, the body parts stay paralyzed. First there must be the intent before beneficial changes can come to the person's body. This is followed by physical effort, which can result in beneficial changes.

So just how does the insubstantial mind or consciousness, this supposed mere projection of the brain, manage to affect the physical brain and body, even cause it to physically alter itself, if it is just a collection of insubstantial thoughts?

The answer is that we just don't know, unless there is some type of quantum basis to the human mind that allows for this. One answer is that human consciousness exists independent of the brain, that it merely uses the brain as an interface to interact with another consciousness and/or perceived reality, just as we might use a telephone to make a text or call to someone else. The brain may only be a medium, according to their theory, and the mind is really independent of, but still reliant on, the physical brain to "get through," as it were. It allows the consciousness to interact with its surroundings.

An interesting theory. So it follows that damage to the brain, and the ability of the consciousness to interact with others is then necessarily limited, but the consciousness itself could be unaffected otherwise. As Wikipedia states:

"It is already known that quantum mechanics plays a role in the brain, since quantum mechanics determines the shapes and properties of molecules like neurotransmitters and proteins, and these molecules affect how the brain works."

And as **Wikipedia** also states:

"The quantum mind or quantum consciousness [1] is a group of hypotheses proposing that classical mechanics cannot explain consciousness. It posits that quantum-mechanical phenomena, such as entanglement and superposition, may play an important part in the brain's function and could explain consciousness.

Assertions that consciousness is somehow quantum-mechanical can overlap with quantum mysticism, a pseudoscientific movement that assigns supernatural characteristics to various quantum phenomena such as nonlocality and the observer effect.[2]

Maybe, these assertions of the nonlocal nature of the mind aren't so "pseudoscientific" after all? Could they explain such paranormal phenomenon as remote viewing (of which there is strong evidence for its existence, even by the C.I.A.) extrasensory perception, etc.?

Chapter Conclusion. The evidence for quantum physics being involved in the human consciousness is mounting. The sense of smell, the way the mind unites separate images stored in the physical brain to complete "a picture," even hearing, and other aspects of our senses may all be involved and reliant upon quantum physics.

And again, at the moment of death (where the brain is still intact at the time and so can be monitored), residual electric energy lingers on, which then fades away, apparently "draining out" through microtubules that actually exist in the human brain and which do seem to operate on a quantum level to some real degree. Some researchers theorize this is the act of the human consciousness leaving the physical body, escaping out into the universe in some way that we simply do not know yet or understand.

Perhaps the human consciousness, mind, or soul, if you prefer, even has weight. A researcher in 1907 claimed this might be so and that it might just weigh about 21 grams. As **Wikipedia** puts it:

*"The **21 grams experiment** refers to a scientific study published in 1907 by Duncan MacDougall, a physician from Haverhill, Massachusetts. MacDougall hypothesized that souls have physical weight, and attempted to measure the mass lost by a human when the soul departed the body. MacDougall attempted to measure the mass change of six patients at the moment of death. One of the six subjects lost three-fourths of an ounce (21.3 grams)."*

So who knows? The information about consciousness, although still growing, is minimal. However, it does seem there is more to consciousness than just the physical brain. The mind interacts with the brain, physically, and this means it is more than just a projection of the physical functions of the brain. Thoughts have an effect on the physical and that is hardly the characteristics of some phantom of existence that many people think the mind might be.

If our minds are just illusions, just products of the brain and nothing more, than how can this be, how can a mere illusion have any impact on reality? It can't. So logic dictates there is more to human consciousness than just an illusion of a self, one brought about by side effects or the result of the physical functions of the brain. Whatever consciousness is, it would seem to be "real." It would seem to be more than just an outcome of the brain itself.

Furthermore, if human consciousness is a quantum-based creation, then might it be immortal, able to exist outside the human body? Remember those microtubules in the brain, that some neurologists think might be the way the human consciousness leaves the body? In the next chapter, we will discuss this concept of life after death in more detail.

CHAPTER 12—Quantum Ghosts and Quantum Souls?

"Not only is the Universe stranger than we think, it is stranger than we can think."

—Werner Heisenberg,
Across the Frontiers

We've discussed the idea that the human mind or consciousness may function, and actually does seem to function, on the quantum level in some very telling ways. Besides saying the human mind must have quantum capabilities in order to exist, there are other things to consider if this theory is true.

Remember the idea that in the quantum realm, a particle can behave like a particle or a wave, depending on whether or not it's being observed? Well there is another aspect to quantum physics with regard to particles. That is, two particles can be "entangled." We don't need to go into great detail on this, but suffice it to say that when particles are created, they come in pairs and these pairs can be entangled where if the "spin" is changed on one particle, then the "spin" of the other particle is affected. Now we aren't concerned here with what "spin" means, but rather the results.

Scientists have found that if the spin of one particle of a paired set is changed, the other particle's spin changes accordingly, and moreover, **THIS CHANGE TO THE OTHER PARTICLE IS INSTANTANEOUS AND NO MATTER THE DISTANCE BETWEEN PARTICLES.**

This is incredible, because it seems to defy Einstein's Theory that nothing can travel faster than the speed of light. Yet, somehow, the information from the one particle is transferred somehow to the other instantly, even if each of the paired particles is at the opposite end of the universe, billions of light years away from each other! Einstein was so upset by this that he referred to this as "spooky action at a distance."

Furthermore, we have tested Einstein's theories of General and Special Relativity and they hold up. The have predicted things that we've later found are true at this point, including the part of the theory about the relative nature of time itself. So Albert Einstein's theories aren't in any doubt. Yet, quantum mechanics is the most thoroughly tested theory to date, and for over a hundred years now, and it, too, has never failed the tests to authenticate the fact if it is true or not.

There is more. Based on some modified versions of the famous "split screen" experiment, where scientists send photons through several narrow slits to then strike a screen beyond, the result has shown that particles can act like waves and vice versa.

There is yet another phenomenon and this one is called "quantum erasure." It seems to defy our idea of the laws concerning time itself. Based on these experiments, it seems by changing the outcome of the experiment, scientists can change the original conditions, the cause, AFTER THE FACT! Let me say this again: it appears as if by changing the outcome, the effect, the researchers can change the initial cause of the effect. How this is even possible is uncertain, but it seems to be so.

As a rough example, it is as if (the cause) you spilled boiling water from a pot onto your arm. You then experience (the effect) an intense and searing pain as a result. However, on the quantum level, in the world of the subatomic, it appears you can experience the effect (the pain) first, and then the "cause" (spilling the boiling water) afterwards. What's more, researchers seem to be able to change the effect and so somehow, on the quantum level can thus change the initial circumstances of the cause, apparently after the fact of it happening.

This means they somehow can change a cause in the past by affecting the outcome, or the effect. It is as if, at the quantum level, at least, that the past can be changed, or at least time can be sort of tricked. This is not only difficult for us to comprehend, but also hard to believe, but on the quantum level, at least, it seems to be so.

One final thing about this; when a particle is in the "wave" state, it has no exact location, but rather is considered to be "non-local." The particle exists as a probability function only, in that it exists in sort of a haze or sphere of probabilities where it could exist as a particle but does not in any particularly precise location. Then, when observed, it collapses into a solid particle and only then has a specific location.

So what does this all mean? Well:

1. Particles, when in the form of waves, are non-local in nature, so instead, they exist as a "probability" wave function that can manifest in only one of a number of probable locations once finally observed, with some locations being more probable for them to manifest in than others. But understand this; although there is much less of a likely probability the particle could manifest somewhere highly improbable, it still conceivably could.

If put in the classical world, on the human scale, and you were a probability wave (actually, you are a collection of them because of all the subatomic particles that make up you), you most probably would manifest right where you are. However, there is a small possibility, very small but still a real one, that you might suddenly pop out into empty space between the stars or elsewhere. Some researchers even wonder if you might not pop out into another time, or even space-time (another reality). This is an exceedingly small possibility, but not being zero, it could theoretically happen, given enough time and the right circumstances.

2. Particles can exist in more than one state at once, being at times in a state of superposition, where they are in two states. Again, this phenomenon is called "superposition" because two states manifest at the same time, one superimposed on the other. I know, a difficult concept, but it seems to be true for the subatomic world.

3. Entangled particles somehow "know" what state their twinned partner is at any moment, as in the example of "spin." If one half of a pair of particles changes its spin here on Earth or anywhere, the twinned particle must change its spin accordingly, and not only instantly, but no matter how far apart they might be, even billions of light years.

4. Quantum erasure would seem to indicate that subatomic particles, at least in the quantum erasure experiment, demonstrate that they can ignore time on the quantum level, at least. The outcome (effect) of an experiment can be altered, thus causing the earlier in time initial conditions (the cause) to be altered, as well. Even though the cause has already happened.

5. For all these reasons, superposition, entanglement, being able to exist as a probability wave, and quantum erasure, particles would seem to have a truly "nonlocal" nature as mentioned earlier, that is space and time seem to not govern them as they do things in our macro world, the world of everyday things around us.

We don't know how this can be, but we know it seems to be so. Experiment after experiment seems to indicate this. And, of course, these explanations are not specific, but simply made here in terms the average, non-scientist researcher can better understand them. The point here is just to get the general idea across to those of us who are not scientists or mathematicians.

What does all this mean for the human consciousness? Well, if it functions on a quantum basis, and evidence seems to show this, then the human consciousness is not governed by the normal laws of classical physics as we've always thought everything to be so governed.

This mean that the human mind might also be "nonlocal." This could mean it might well be able to exist outside of the human body (dead or alive), and thus result in experiences such as so-called astral project and out-of-body experiences.

This nonlocal ability of the mind would then also explain near death experiences, as well. Remote viewing would have an explanation, that is, the ability to see things with one's mind that are far away and so not in one's normal sight range. Nonlocal behavior of the consciousness could explain much else, such as the idea that the human mind could perceive ghosts. Perhaps they act or are a disembodied human consciousness that sometimes can manifest itself in the material world, even as the human mind can make changes in the physical brain.

One more thing it would explain is reincarnation. If the human mind is not subject to the limitations of linear time (moving from the past to the present to the future), then perhaps it can reincarnate, but more on that later.

The Human Mind Does Not Seem to Be Confined to Only the Present. This may be hard to believe, but it would seem to be true. Numerous experiments, repeated many times at many mainstream universities have shown that the human consciousness exists not only in the present, but also partly in the future, as well.

When participants in the experiments were shown a series of randomly generated images, most positive images, but with the occasional horrific or negative one, as well, their bodies reacted to the horrible image before it was shown. This was done over and over, and proven repeatedly at different universities.

The average time a person might respond to a coming terrible image ranged anywhere from 2 to 10 seconds, so the researchers have come to the conclusion that the average is (of course) right around five seconds for most of us. They don't know how this can be or why, except to theorize that it might have evolved as a means of survival. If our ancestors felt something was about to attack them seconds before it did, it might just have given them enough time to respond to save themselves, so better chances of survival. They would then live to breed and so continue this ability in our human population. This last is just a theory, though, and nothing more.

Also, the human consciousness lives partly in the past. For example, by the time the human brain registers something it has seen, 13 milliseconds or more have passed. Then it takes even more time for the human brain to physically respond. There is accumulated lag time, as it were. This is just based on physics, and the limits of the body, brain, and the speed of light.

Now, this may not seem like a long time, but the faster something is happening, the more this has to come into account. For instance, if the pitcher in a baseball game throws a fastball at 90 miles an hour to the batter, that batter when he swings his bat isn't swinging at where he sees the ball at that instant, but at where his mind predicts the ball will have traveled to in the intermediate time. Remember, there is that, at least 13 millisecond of lag before the eye registers the ball (in which time fast ball has moved on), and then add to this the physical lag time in the brain ordering the batter's muscles to respond and swing at the ball.

Yet, somehow, the batter still manages to hit the ball. The same holds true for race car drivers. They are speeding along, some have even gone at incredible speeds, and that 13 milliseconds means their car has already moved beyond the point where the mind has just registered it to be. Yet somehow, they seem to compensate for this and still manage to drive their vehicle, make turns exactly when needed, and brake the right distances when necessary, if not always perfectly....

Chapter Conclusion. We have seen in this and the prior chapter that the mind does seem to function in quantum ways. And subatomic particles in the quantum world can exist in a "state of probability," as well as exhibiting nonlocal behavior, being neither subject to distance (space) or time, as in the experiments with quantum erasure seem to show. If this is so, and the mind is quantum in nature, then it, too, could well be nonlocal in nature, with regard to not only distance, but time, as well.

Again, this could account for out-of-body experiences (OBEs) and near death experiences (NDEs). It could also supply an answer to how reincarnation might work, or even the existence of what we think of as ghosts. The quantum mind may not be subject to classical physics as we know it. Rather, it may merely use physical human brains to manifest itself to be able to interact with others.

Again, damage the brain, and you have damaged the medium (transmitter and receiver?) for the mind, but you have not necessarily damaged the mind itself. Does the brain have a strong influence on the mind? Most assuredly. Does the mind, that ethereal thing composed of only thoughts, ideas, and dreams have an influence on the physical brain? Also, most assuredly it does!

The human mind is more than just a function of the physical brain it would seem. It is not just a projection on a screen created by the projector, the human brain, but rather seems to be "something else," and "something more." What's more, it may not be limited to this one reality, as we'll see in the next chapter.

CHAPTER 13—Quantum Suicide or Immortality

"Quantum physics teaches us that we can simultaneously exist in many places, under certain conditions."

—Amit Ray,
Quantum Computing Algorithms for Artificial Intelligence

    We come to yet another way we might have immortality, the one that is mentioned in the title of this book. Besides having multiple versions of ourselves, perhaps even an infinite number of us and possibly in a number of variations, as well, there is another way we might consider ourselves immortal. This is a more direct one, where the self continues, and it, too, is based on the Many Worlds Theory brought about by quantum physics.

    No, I don't mean that if you die here, another version exists of you somewhere or some when else, although, those scientists who hold with the Many Worlds Theory think this is probably true. Rather, this is the idea of quantum suicide, or quantum immortality as it is also known, which started as a simple thought experiment.

    The principle is based on the dual nature of quantum physics (wave/particle phenomenon). For many quantum physicists, they accept the many worlds interpretation as the basis for the idea that a particle can exist in more than one state, either as a particle or wave, but the idea goes further It also states that when a probability wave collapses into a "real" particle in one reality, that is also stays as a probability wave in an offshoot reality that branches the moment the choice is made to observe it.

    In other words, every time a decision is made, as in someone deciding to observe a particle, a parallel reality comes into being where they did not make that choice. So rather than Schrodinger's cat as in the famous experiment where the cat is in a box and may at any moment be killed by the release of a poison if a subatomic particle chooses to decay. In this thought experiment, until the box is actually opened, the cat is said to be in a superimposed state where it is both alive and dead at the same time.

    However, some researchers argue with the Many Worlds Theory that the cat doesn't live in two distinct states of being that collapses to just one, alive or dead, when the box is opened. Instead, when the choice is made to open the box, whether the cat is alive or dead in this universe, immediately there is another parallel world created where the cat is in the opposite state. If you open the box and the cat is alive in this universe, there is an immediate creation of a parallel universe where it did not survive and you found it dead. This is pretty basic as an explanation but it does suffice for our purposes here.

So how does this idea compare with humanity? Well if a person dies in this reality, there is immediately created another reality where the person did not die but continues to live. A split in reality forms for the person, a superimposed state where in one reality they died, but in another they go on existing. For the person involved, there is no interruption in the continuity of their existence from their point of view. They don't perceive that they died at all, because that version of them no longer would exist, but rather they would just go on living in the alternate reality, their life seemingly uninterrupted.

So it is only the person's continuity, uninterrupted existence that goes on. For the observers in this universe, however, the individuals have died. But for that individual involved, all that matters, is that they go on living, and they are unaware the universe split at that exact moment of their death, because for them, it simply didn't happen.

This thought experiment was originally created by Max Tegmark and he referred to it as "quantum suicide." He based the thought experiment on the idea that if one fires a bullet at one's self, (a quantum subatomic particle bullet in his version, but only if the particle has a downward spin to it, and not upward). From an observer's perspective, both outcomes have an equal chance. If the spin of the particle is "upward," the bullet doesn't fire. If it has a "downward spin" it does and the person then dies. But for the victim, to have continuity of consciousness, the gun never fires, or at least the bullet doesn't, although statistically, according to the laws of probability, it often should.

For the victim then, they were never shot and they continue to go on existing, unaware that in a parallel reality that formed, they died. For the observer, every now and then the gun will fire and the person will be dead. But again, for the victim, life goes on uninterrupted. Reality has simply split and their lives go on in one, as does their consciousness. Since it is only a conscious being that can be aware, the split in reality continues on with the version that is still capable of being so aware.

Physicist David Deutsch, who is an adherent of the many-worlds interpretation, states:

"regarding quantum suicide that it would not work under the normal probability rules of quantum mechanics. Instead, one would need to add an additional assumption of ignoring worlds where the experimenter is not there. He believes that assumption is false. [5]

Physicist David Wallace said: *"...that a decision theory analysis shows that a person who prefers certain life to certain death must prefer to keep themselves alive in worlds that are more likely outcomes, not just in less likely ones. [2]"*

However, Physicist Sean M. Carroll, who also likes the many-worlds interpretation, states:

"...about quantum suicide that neither experiences nor rewards should be thought of as being shared between future versions of oneself, because these future versions become distinct persons when the world splits.

He then states" *...that a person cannot pick out some future versions of oneself as really being oneself and not the others. He concludes that quantum suicide kills some of these future selves, which is a bad thing, the same as if there were no other worlds .[6]"*

So one can see that there are arguments over the quantum immortality idea. And as involved as this sounds, it simply is a series of variations of the idea of the Many Worlds Theory. That every time there is a choice/decision made, even on the tiniest level, that both outcomes must occur, and the universe splits at the moment of decision, each reality expressing a different outcome, and so for a consciousness with regard to dying, at the moment when fate or circumstances brings about apparent death, reality splits and that person goes on living as if nothing ever happened in a parallel existence. And indeed, for that person, there was no break in the apparent continuity of their existence. They did not breathe their last, but simply took yet another breath!

Chapter Conclusion: The idea that the universe splits every time any sort of decision is made, such as whether an atom splits or not, may seem incredible. And you are right, it is. But many mainstream researchers believe this might explain the dual nature of particles as being both a wave and a particle until actually observed. And notice the probability wave of the particle, as to where it manifests itself is first dependent on have been observed. So many scientists think that all of the probabilities of that particle's wave state must be realized somehow, and this, they fell could well be by being manifested in parallel universes, even as outcomes of life and death must also be.

One way it to think of it as being on a train with man possible branch lines. The instant the train crashes on one line, killing you, another branch line pops into existence where the train didn't crash and so you go on living, your life uninterrupted, and the crash never seeming to have happened for you. This is the idea behind quantum immortality. And weirdly, it may be true. Again, a great movie that plays with this idea, and there are several, is *"The Discovery"* with Robert Redford, but there is another, high action-packed one that also deals with this idea in a truly brilliant way, as well, Called *"Source Code,"* Starring: Jake Gyllenhaal and Michelle Monaghan among others. The former movie deals with the idea of quantum suicide, whereas the other deals with quantum immortality in a more factual way. You might want to consider watching them.

CHAPTER 14—Are All Probabilities Carried Out?

With regard to the Many Worlds Theory, there is one more aspect of it to consider and we only briefly touched on this in the prior chapter, and that is the likelihood that all probabilities are carried out all the time and in every possible way. There is some evidence, fairly good evidence, that this may, in fact, be the case. Remember earlier when we were talking about parallel realities where all probabilities may play out for something, such as the wave versus particle state of a subatomic particle? Well, let's take a closer look at this idea because it has ramifications for immortality.

To start with, we have to understand two things: (1) a probability means it is actually something that is capable of occurring at all, however small the chance of it actually happening might be. In other words, the probability of something happening can be almost infinitesimally small, but it can't be zero. There has to be some chance, however small, that something can actually occur. (2) the second thing one has to have is that this must work for us on the classical physical level, the everyday world and not just in the world of the subatomic.

For those who took any science in school and had to study the concept of atoms at all, the standard diagram/image we were given was that an atom is composed of a nucleus with an electron circling it. The image usually always provided was of something like the Earth orbiting the sun. This, unfortunately was incorrect. Although, it was meant to simplify things and so better help students to understand the nature of atoms, it was just plain wrong.

Electrons do not circle the nucleus as the Earth circles the sun. Instead, they surround the nucleus in a sphere, a rather fuzzy one at that. This is because they do not occupy any one position at any given moment. Being in the wave state, unless observed (measured). The electron is just in any given spot around the nucleus in a sphere as a set of probabilities, everywhere in that sphere encompassing the nucleus of the atom, but nowhere in particular. The electron, in this state, functions as a wave, in this case, a wave of probabilities as to where it might actually manifest if observed.

This is true for all subatomic particles. Mind you, there is normally no chance the electron can exist anywhere outside of that fuzzy cloud or sphere, so that's why it is best visualized as a fuzzy sphere, rather than like a planet orbiting a sun.

Since atoms are made up of subatomic particles, and molecules are made of those, atoms, this means that all macro things, things on the large scale, ourselves included, are really just a compilation of these superimposed states/probability waves. We appear solid to ourselves, even as everything around us does, but reality may be that everything around us is really just a sea of probabilities, waiting to collapse whenever observed, and a decision is made to do so, and only then. It may be that the only thing that determines reality is consciousness from instant to instant and that our consciousness just otherwise exists in that ocean of probabilities, that fuzzy cloud of possible outcomes.

The question then often arises that if this is so, if all is just probability waves collapsing from instant to instant, is the universe just an illusion, and if it is not, how come it doesn't keep collapsing into different probabilities, perhaps even completely different types of universes? Well, those proponents of the probability wave idea of the universe say that the universe keeps collapsing from instant to instant from a wave state to a "particle" (or as we think of it real and solid, so to speak) to its most probable state, which is this one we are in now.

Could it conceivably collapse to another probability? Yes, theoretically, but if consciousness dictates the universe's collapse from the wave state to the solid state we see, it is also manifested as a result of consciousness. It could be that consciousness "expects" this particular universe, or simply that this is really the most probable version of the universe. Some researchers even suggest that it is this instant-by-instant collapse of probabilities to reality that gives us the feel of time passing. An interesting idea, at the very least, whether correct or not.

Chapter Conclusion. If this idea is so, then we, or at least our consciousness, is "outside" of that sea or at least separate from it, so our consciousness may not be confined to and limited to the human body we inhabit.

This means that from instant to instant, we are "choosing" with each decision which reality we are living in. In other words, we, or at least our consciousness, could well be hopping from one alternate reality to another all the time based from moment to moment on the decisions we make.

Another thing it might mean is that our consciousness may only be using our body as a form, a means of interacting with reality around us. If our consciousness is truly defined by this idea, then it would also mean, again, that our consciousness may be able to exist separate from the body, may be able to move about in space and time, as remote viewers claim, and as with researchers who have determined the human mind may exist partly in the future, the past, as well as the present. We may be nonlocal in both space and time.

Furthermore, many researchers think, based on the Many World Theory, that this approach to reality might just be true, or at least they consider it a real possibility. The universe may just be a mass of probability waves with an incredible, if not quite infinite, number of outcomes.

CHAPTER 15—Evidence in The Body for Life After Death?

"In sorrow we must go, but not in despair. Behold! we are not bound for ever to the circles of the world, and beyond them is more than memory."

—J.R.R. Tolkien

Remote, OBEs, NDEs, all might be connected and indicative of something truly profound. Our bodies, themselves, may hold evidence of our abilities for our consciousness to survive death. Remember, scientists have discovered microtubules in our brain that a number of neurologists believe seem to be able to act and behave on the quantum level.

If this is so, then our consciousness may arise up from the quantum realm, and thus manifest through those microtubules into our brains. One neurophysiologist even went so far as to say something extraordinary. Noting that there was residual electrical activity in the brain after death had occurred, he theorized that was the human consciousness draining out through the microtubules to the universe at large, or perhaps into the quantum realm, which may be the same thing, since at the quantum realm things seem to be nonlocal with regard to space-time?

If this idea seems incredible, one must take into account the thousands of reported cases of NDE's and OBE's (Near Death Experiences and Out of Body Experiences). Many people have reported that they have experienced life after death of a sort. When some people have died on the operation table, when their heart had stopped beating for instance, they experience their consciousness leaving their bodies.

This has happened so many times, that there is no doubt it is a real phenomenon. Whether it is a subjective one, or an actual one, remains to be seen. However, since so many of these cases are remarkably similar, and some people are able to describe what was happening around them when they were supposed to be dead, there does seem evidence to suppose it might be an actual event that takes place under such circumstances. If true, then this means the consciousness can survive the human body.

Besides this, there are what are called OBEs. These too, seem to be a genuine experience, and even can revolve around other parapsychological phenomena such as farseeing, or otherwise known as remote viewing.

The person is still inhabiting their body, apparently, but is able to project their consciousness outside of it. The CIA has investigated remote viewing and found it has shown some real success. Once more, this means the human consciousness may not be restricted to the physical body, but may be able to free range. Other evidence with regard to how the human consciousness works with regard to time, also seems to point to this fact. Some people claim, with some evidence to support them, that they can remote view events not in the present.

If so, then we do have real evidence for the idea that the human consciousness can leave the body, or at least can exist outside of the human body. This in itself is amazing, but the implications for immortality are just as amazing. Does the human consciousness totally depend on the human brain? Or can it survive the death of the physical brain? These are all questions waiting to be answered. Yet there is tantalizing evidence that consciousness could well survive the death of the brain and continue to exist.

Chapter Conclusion. If the mind can continue to exist outside of the body, does that mean it can be in a sort of "ghost" state? Is this what people are actually seeing when they claim to see a ghost, a disembodied consciousness? And if the consciousness is not subject to the restriction of time, is this why some people claim to see ghosts or past ghost-like events, such as sightings seen at historical places like Gettysburg? This idea might also account for how some "ghosts" or disembodied consciousness' might be able to actively interact with the living, as well. All this is conjecture, of course, but still, it is intriguing stuff. One thing about the idea of the quantum nature of human consciousness is that it could explain a whole host of such supposedly supernatural things.

PART 4—THE HYPERSOUL

CHAPTER 16—The Hypersoul

"We cannot be born, and we cannot die. Like all energy, we can only change shapes and sizes and manifestations."

— John Green

The hypersoul is a term I have coined to describe what the true nature of consciousness or the soul might entail. The term refers not just to a sort of supersoul that may exist if all things in the universe are connected in a field of energy which may somehow be interconnected, as some famous scientists and researchers believe, but instead, hypersoul goes farther. A supersoul would be a soul connected to others on some level in this universe, perhaps via the Higgs Boson field that scientists claim permeates our entire universe, and would exist as through in a collective subconscious, as Carl Jung believed

Alternatively, the supersoul might be capable of being somehow having every consciousness "recorded" somehow in something like the Higgs Boson Field and as some might refer to it as the Akashic Records of ancient legend.

The collective unconscious, as Carl Jung viewed it, was the ability of the human mind to access a fundamental level of unconsciousness. This level would be a collective unconscious of everyone, one that is one shared by all humans and so could be tapped into at some level by anybody. Jung pointed out that people all seemed to have archetypal symbols in their dreams, and even Sigmund Freud believed that last part as well, and these were symbols from the collective unconscious and so the same for everyone.

Water could be the symbol for fertility, while palm trees, snakes, and skyscrapers could be symbols of the penis, even as gates, doorways, etc., could be archetypal symbols for the vagina. One major definition of archetype, is defined as:

"the original pattern or model from which all things of the same kind are copied or on which they are based; a model or first form; prototype". –Dictionary.com.

He believed these can be added to and/or even modified by a collectively human shared subconscious, as when many people began dreaming about atomic explosions after the advent of the Nuclear Age Post-

World War II. Atomic explosions could well have been the symbol for something explosive, catastrophic, or just a major change in the person's life. This shared symbolism was possible because the collective unconscious was just that, a shared level of unconsciousness between all humans.

If this is truly so, then on some level, we are all interconnected in this way, forming an extended or supersoul in this universe. However, if the multiverse exists, and many scientists think it must, then there must be a multitude, perhaps even an infinity of parallel universes or worlds, with many being almost exactly like ours, even to having "us" in them. If this is so, then it might just be that something as disembodied as the human mind or soul would seem to be, might just be connected on some level with all those other versions of ourselves. So instead of just being a supersoul, one would have to refer to it as a hypersoul, one that stretched across the multiverse.

Sound farfetched? Perhaps, but there are even recordings you can buy, as well as free meditations you can listen/view on the Internet (YouTube, for example, has a number of free ones), that try to teach through self-hypnosis, the ability for the individual to "quantum jump" to parallel realties, in order to interact and even learn things from our mirror counterparts that might exist there.

If such realities exist, and they seem likely according to a host of researchers and scientists from such varying fields as cosmology and quantum physics among others, then it is conceivable that the human mind, at some basic level, unconscious or subconscious perhaps, can, indeed, interact with our other versions of ourselves. This, then, would from a sort of hypersoul, one that transcends the multiverse and not just our universe. Michael Talbot, in his book, *The Holographic Universe*, a bestseller on the New York Times Bestsellers List for over six months, talked about how the whole universe is interconnected, bathed in light and other forms of energy from one end to the other. It was his premise that nothing, therefore, is isolated, that all is interconnected in this fast field of energy which permeates the entire universe including the human mind.

We know that the universe is a sea of energy, even on the most basic levels, and that not just light, but other forms of energy envelope everything in the universe, with the Higgs Boson Field being the most recent discovery in this regard. So is it such a stretch of the imagination to think that this all-encompassing sea of energy might even extend to parallel realities, as well?

If so, what is to stop the quantum nature of the human mind from interacting with other versions of ourselves in those parallel realities? Our quantum physicists have already made it clear that time and reality, as we think of them, behave much differently on the quantum level. It may be that the quantum realm does not stop at the boundaries of our own universe (and some scientists think this is so) but actually might just form the basis of all possible realities. Remember that sea of probability waves where all outcomes must happen? Some researchers are convinced that all probabilities, all possibilities must play out in such a way, and if not in this universe, then in parallel ones, as well.

Chapter Conclusion. Hence we have my idea of the hypersoul, that the mind or soul, quite possibly being of a quantum nature, might also exist in multiple universes, the multiverse, as it were. This would mean that human consciousness is not just limited to this particular universe but extends across all of space times and all universes that we might have other versions of ourselves. Perhaps this is where the feeling of being "at one with the universe" might come from?

Is quantum jumping really possible? The answer to that question still has not been made clear, but it if is possible, or perhaps even probable, and one can on some level interact with other versions of ourselves, then the hypersoul definitely would exist under such conditions and perhaps does, even if quantum jumping in and of itself does not.

This in turn would form a sort of cosmic oneness for us on a truly vast scale, one beyond our limited abilities to understand as we now view reality and even our particular universe. So when many people experience the near-death experience of "going towards the light," just maybe they are experiencing the approach to that vast cosmic oneness, that infinite sea of interconnectedness of the self with all the truly countless other versions of ourselves. Food for thought?

CHAPTER 17—A Conscious Universe?

"We are the cosmos made conscious and life is the means by which the universe understands itself."

—*Brian Cox*

The idea that the universe might be conscious is a fascinating one, and does seem to have some points in its favor, scientifically speaking. Some even go so far as to say that human consciousness, along with all forms of consciousness, is the universe trying to comprehend itself, that it is being conscious through all of us. Others declare that if the universe is full of conscious beings, whether us and/or others, then the universe, in a very real sense, is conscious. This idea would have to have several things to support it to be true:

1. There would have to be life in the universe, perhaps more than just here on Earth, although even that would fit the need. Since we are alive, do exist, this first requirement is fulfilled.

2. This life should be capable of being conscious in some sense, as we humans are and perhaps, animals to a lesser extent. Therefore, the second requirement is also fulfilled.

3. The universe should be able to act in a conscious way to some degree as a result, if only through us and our self-awareness and the ability to comprehend. This final necessity is also fulfilled. Humans are, of course, self-aware and we have taken great strides in learning about our universe, even if we still have much to learn in that regard.

So under these conditions, the universe is conscious because there is life, at least on earth, and conscious life in the form of animals and even more so, self-aware life for sure, in the form of us humans, at least. Latest news also suggests there might be life in the upper clouds of Venus, microbial life. If this is so, then we might have life in more places than just Earth, which further supports the idea the Universe may be conscious, since life of some sort may exist in other places, and even possibly intelligent life.

Given all this, how does it affect the idea that we might be immortal, or that there is life after death for humans? Well, unfortunately, under these circumstances, it probably would not guarantee such a thing. If the universe is only conscious through life that exists within it, then life acts more like an organ of the universe, rather than some vast, disembodied, but "whole" entity in its own right. Individual Brain cells in humans die, for example but the human mind can still persist. Perhaps, we as individuals are like brain cells and so have no hope of an eternal life if this idea holds true?

If the universe is conscious only through isolated and separate living creatures, then those individual creatures are not connected, apparently, and therefore, will live and die and that will probably be that. Again, it is rather like a cell in the human body dying. The body goes on, and the conscious mind with it, but the cell is gone, presumably without the consciousness even being aware of it having existed on an individual level.

This would be a sad state of affairs for the idea of immortality for humanity, if so. The entire human race, even as already have the dinosaurs, could become extinct and the impact on the consciousness of the universe would be minimal (unless we are the only species in the universe capable of contemplating the universe). It would rather be like that single cell in the human body dying, rather than the brain as a whole.

However, there is another approach to the idea of a cosmic consciousness, and this is intriguing because some researchers theorize that there is such a universal consciousness, that it is aware, and it is all interconnected throughout the universe. How could this be? Well:

1. The universe may be more than the sum total of its living parts, even as the human mind seems to be more than the sum total of all the parts of the brain and body combined.

2. The universe does seem to be permeated by vast fields of energy. These, such as the Higgs Field (Higgs-Boson) and even light itself, fill the entire universe, as mentioned earlier in this book. These could then act as a sort of carrier wave for consciousness, or at least a medium that could carry consciousness within it throughout the entirety of the universe.

Combined with the idea of the nonlocal nature of particles/waves, that something 5 billion light years away, for example, can be affected instantaneously by something happening here, as with paired particles, then we could have a consciousness that might span the entire universe. In fact, even the arrangement of galaxies, the way they are spread out in interconnecting webs, oddly mirrors the neural nets inside the human brain.

The cosmic web and the human neurons or neural network are amazingly similar in look and design as shown in the photographs below.

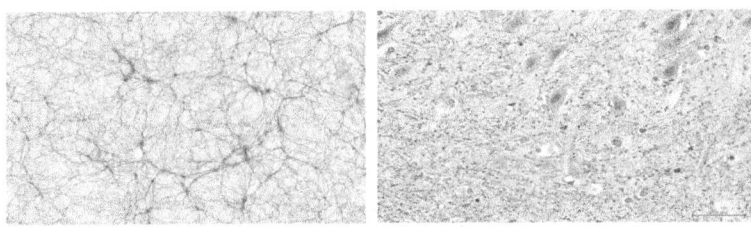

Courtesy of Universe Today,
https://www.universetoday.com/148966/one-of-these-pictures-is-the-brain-the-other-is-the-universe-can-you-tell-which-is-which/

"One of These Pictures Is the Brain, the Other is the Universe. Can You Tell Which is Which?"

Panpsychism. Along with the above idea, and rather akin to it in many respects is yet another way the universe may be conscious, and this is called Panpsychism, with "pan," the prefix of that word meaning "all" or "everything" as two of its main definitions, and of course, in this case with "psychism" from the *Century Dictionary* meaning:

"...the doctrine that there is a fluid diffused throughout all nature, animating equally all living and organized beings, and that the difference which appears in their actions comes of their particular organization."

Now Mr. Phillip Goff bases his theory of panpsychicism on a slight variation of this idea of a "fluid" permeating all nature, in that he believes consciousness arises from the *"consciousness* [that] *pervades the universe and is a fundamental feature of it."* So we humans would be a physical manifestation of that all-pervasive consciousness, rather like a finger is an extension of the hand.

Whereas, scientists currently consider that consciousness as we know it, as strictly a product of the brain whether quantum based in its nature or otherwise, and that the more advanced the brain, the more consciousness or self-awareness there is, Panpsychism instead believes that all things are conscious, at least in a way. That is not to say that a subatomic particle is self-aware or even a rock or piece of metal. However, they may have a primitive form of consciousness, being part of the universe.

No, Mr. Goff feels that just as a mouse is more aware than an earthworm and a human is more aware than a mouse, so the same holds true for the universe, itself. And by consciousness he doesn't necessarily mean self-awareness as humans have it, but rather simply the ability to experience things, as such basic things as pain, or perhaps that which "feels good" on its most basic level. For example, we're not sure how self-aware cats are. We are fairly certain however that they do "enjoy" having their ears rubbed, or their back and chin scratched. So they are capable of experiencing feeling good.

Therefore, on the level of the universe, this, too, might hold true. Where a subatomic particle, such as a proton or neutron is very limited in the extreme in its consciousness, the more complex a system is, the more conscious it may be or may become. Just as the human brain seems more conscious than, say, a tree, yet a tree would have some level of consciousness because of the complexity of the systems which allow it to function and so exist.

On that basis, if we take the universe as a whole, it is an extremely complex system, certainly as complex if not much more so than the human brain if one takes into account the particles that make up such systems, which in turn are part of even larger and more complex systems, and so on. Moreover, if the universe is infinite, then its level of complexity is mindboggling in nature. The universe with its planets, stars, galaxies, cluster super clusters, and so on makes for a very complex system. If carried to its logical conclusion, the universe maybe "superconscious" to the point of being what many might consider as God or a godlike being.

As metaphysical as this approach sounds, it does account for not only what we experience in our minds but perhaps even more importantly *how we experience.* Where science is mathematically based and so quantitative in nature, Mr. Goff's approach would help us solve how we experience what we do. It is all well and good to say the human eye perceives photons, which then are sent as a signal to travel up the optical nerve to the brain, but how do we then create images of what those photons, just packets of energy are really, into three-dimensional images of what we are looking at. At this point, we know the mind recreates these images, and not always exactly as the original was, as with our minds being tricked by optical illusions when our mind persists in seeing something that doesn't actually exist "out there."

How is it that we can smell so many different odors when our olfactory senses are just responding to molecules of matter? How can we then have such strong emotional responses to such things as the smell of a flower, or the view of a sunset over a lake? Quantitative, mathematical-based scientific approaches can determine the physical mechanisms of how signals reach our brain, but they, so far, have been woefully inadequate in determining how we can recreate images of the world around us inside our minds, not only visually, but replete with sounds, smells, physical sensations and even react emotionally to them.

How is it we can appreciate and enjoy a tune, for example? Shouldn't melodies just be a series of disconnected notes, ones following each other in a disconnected sequence? How is it our minds can create a continuity of sounds in order to appreciate such things as melodies? Remember, music is terribly reliant upon time. The human mind must remember the last note played, while simultaneously hearing the one presently being played, and also still be able to anticipate what the next note will be in order to comprehend a melody.

Panpsychism instead would explain this ability with the idea that consciousness simply exists naturally in the and throughout the universe.

Chapter Conclusion. One thing all these concepts of a conscious universe seem to have in them or strongly related to them is that consciousness is something that occurs naturally and seems to increase in ability (more self-awareness) in relation to how complex any system is. When one considers the universe is really all energy, that even matter is sort of a form of condensed energy as Einstein proved with his famous equation $E=MC^2$ where energy is equal to mass times the constant (speed of light) squared. Rearranging this formula one can convert not only mass to energy as here, but also energy to mass. Scientists favor the concept of String Theory as an explanation of things, and this theory presupposes that when all matter is reduced to its elemental particles, these are really just loops of energy.

Therefore, if everything is just really energy in its essence, then when Walt Whitman wrote *I Sing the Body Electric,* he may have been right on the mark! If all is really energy, then ultimately, consciousness and self-awareness somehow, must be a function of that energy. And since it must be, then perhaps there is a "cosmic all" and at the base of all reality is a "cosmic oneness." Moreover, if consciousness is a function of energy than it is not just limited to "here" or "there" but may be everywhere at once, or at least capable of manifesting anywhere, and possibly even any when?

If this is indeed, so, then we are not separate parts, isolated and distinct from each other and the rest of the universe because Panpsychism incorporates the idea that everything is part of the whole. As Walt Whitman's closing line of that same poem says:

*"O I say these are not the parts and poems of the body only, but of the soul,
O I say now these are the soul!"*

As New Age as such an approach sounds, it is a logical progression based on Mr. Goff's initial premise. In which case, we are not lonely, separate, and tiny entities that are insignificant in nature, but rather we would be part and parcel of the cosmic consciousness, the "cosmic all" as so many philosophers have mentioned throughout history. Perhaps we are even part of one grand cosmic supersoul, or hypersoul as mentioned earlier in this book. Who knows? But if we arose from such a cosmic consciousness, perhaps our minds return to it upon the death of the body. Perhaps, just perhaps….

CHAPTER 18—Evidence for Near Death Experiences?

This is necessarily a short chapter since most readers are already experienced in the idea of near death experiences and out of body experiences, so this section simply deals with the matter in a more passing manner, and is included for those who might not be so well-versed in the subject. That it is mentioned here is primarily because of the connotations the topic has with regard to the idea of the consciousness being able to outlive the body. And it does seem to further illustrate and provide some evidence, if only marginally so far, for the idea that the consciousness might be able to survive intact beyond the body, whether the body is still living, is dying, or is clinically considered dead for a time.

Numerous anecdotal accounts of near death experiences, where a person who is clinically dead feels as if they are floating up out of their bodies and can still see and hear everything going on around them have been reported. People also report tunnels leading to a white light, being reunited with loved ones, or some sort of loving being of light, etc., coming to them.

As stories by themselves, these would, at first, appear have little scientific credence, and many medical researchers do dismiss then as just hallucinations of a dying brain that is being deprived of oxygen. However, other researchers are not so quick to dismiss the idea of near death experiences or NDEs as they are often referred to. Again, to put it bluntly, most scientists and even researchers still adhere to the idea that near death experiences are nothing more than hallucinations of a dying brain, perhaps because it is denied oxygen or is simply shutting down.

However, there are those researchers and doctors that take issue with this conclusion, especially one prominent neurosurgeon who once believed the above explanations were the very cause of NDEs. His name is Eben Alexander III and he, of late. has become an author. This is for a very good reason, because he, himself, experienced an NDE.

In his book, *Proof of Heaven: A Neurosurgeon's Journey into the Afterlife,* which came out in 2012, he described his NDE when he fell into a coma, a medically induced one in 2008. This was as a result of him acquiring meningitis, which almost resulted in his death. He describes, vividly, his near death experience, and now he is a convert to the idea that NDEs exist, they are real, and not mere hallucinations. This book was a bestseller, in large part because of his impeccable credentials as a neurosurgeon.

Other doctors and nurses have reported similar occurrences with various patients whose hearts have stopped on the operating table, for instance or even in other life threatening situations. When the patient(s) revived later on, they often reported NDES with varying details, but all accounts seeming to possess major commonalities, as well. Patients report a tunnel-like effect, floating up out of their bodies, seeing and hearing what's going on the operating room while they are technically dead, having dead loved ones come to them to help guide them on, and also, very often, a bright light, as well as a sense of pure joy and peace, among other things. In fact, many patients when told by ethereal beings such as dead relatives that they must go back into their bodies, they resist, preferring to stay where they are.

Again, one could dismiss these out of hand, except for a couple of things; one in ten patients who "die" in this manner report NDEs according to one source. Another source says the number is higher, being 17 percent of patients rather than 10 percent. Either way, that's a lot! According to Jeffrey Long, MD, in his article, *Near-Death Experiences Evidence for Their Reality,* in the Journal of the Missouri State Medical Association:

"While no two NDEs are the same, there are characteristic features that are commonly observed in NDEs. These characteristics include a perception of seeing and hearing apart from the physical body, passing into or through a tunnel, encountering a mystical light, intense and generally positive emotions, a review of part or all of their prior life experiences, encountering deceased loved ones, and a choice to return to their earthly life."

Also, the descriptions some patients give of what took place in the room while their hearts were stopped and they were clinically dead were so accurate as to amaze the nurses and doctor(s) who were in attendance when the event occurred. Those in attendance simply couldn't account for how a supposedly nonresponsive patient, one whose heart had stopped, could accurately tell them what had gone on, even to what was said by those doctors and nurses in attendance at the time of the event. From everything that medical science knows, such things should be impossible.

Moreover, people who have out of body experience's, OBEs, can often describe in detail such things as events and places outside of the operating room, and can often do so accurately. Again, these can occur outside the room, in hallways, outside the building, and other places removed from the operating room, when they couldn't possibly have known about them, being supposedly dead, but which they still then could accurately describe. Some of these reports seem almost more like OBEs (out of body experiences), rather than NDEs, as a result.

The only real difference between OBEs and NDEs seems to be that when someone has an OBE, they do not meet dead relatives or friends, nor do they go through a tunnel toward the oneness of a bright light. Instead, although they float out of their bodies, and often can see their sleeping bodies below them, they seem restricted more to this reality than that described as an afterlife, although some claim that space and time have no meaning. Some claim they can travel through space and even to other universes and/or times to witness events.

Chapter Conclusion. Given these accounts of both NDES, and some OBEs, it seems there is tantalizing evidence that the consciousness can leave the physical body, both while living and when the body dies. After all, if even a neurosurgeon can believe that a near death experience really happened to him, one who was opposed to the idea prior to his own experience can be converted to believing in them, then just maybe, NDEs and OBEs might be real. This would provide further evidence for the idea of the consciousness/soul being immortal. Further investigation of this matter is necessary to reach any real conclusions, but there are doctors, nurses, some researchers and even one neurosurgeon, Dr. Eben Alexander, who are convinced the phenomenon is real.

CHAPTER 19—Evidence for Reincarnation?

"If we've been born once already (which we know we have) why then is it so hard for some to believe that we've been born before? The answer to that is nothing other than the information about life one has previously received."

—Renee Chae,
This Thing Called Life: Living Your Ultimate Truth

"Consciousness is endless, from one incarnation to the next. It simply will and does manifest in other places and times, regardless of what becomes of the human race."

—Zeena Schreck, Beatdom
#11: The Nature Issue

Now we follow on the idea of Mr. Goff's Panpsychism. This is the idea that if everything is a result, function, or manifestation of energy, and since energy, as with matter, cannot be created or destroyed but rather, merely transformed according to the laws of physics, then the concept of reincarnation isn't necessarily so "out there" after all.

If the mind is energy of a form, and if it can exist in the fields of energy that permeate the entire universe (as in the form of light, Higgs Boson Field, etc.) then what's to stop that mind from coalescing in other bodies? In other words, reincarnation, that is of the soul, personality or mind of an individual in successive physical bodies?

The idea of reincarnation is tantalizing as is the evidence for it. Might we have immortality through reincarnation, being born again and again? Is there any real evidence to support this idea? Perhaps one of the oldest questions humans have dealt with is the idea of reincarnation.

The term literally means, "the taking on of flesh again." However, as humanity developed, just how this process could have occurred took on different variations. And of course, within the last couple of thousand years the divide has only grown greater, especially between the east and the west with regard to religions and philosophy.

Whereas the "East" has developed religions that encompass the idea of reincarnation, this hasn't been nearly the case in the "West." Faiths that embraced the concept in Asia include Buddhism, Hinduism, and even Jainism, whereas Christianity and the Muslim religions of the Middle East and the West veered away from the idea almost in its entirety. This is not to say that no one in the West believes in reincarnation. There are various movements and sects that do.

For Hindus, the idea of reincarnation is the belief that the soul after physical death, begins anew in a new body, whether human or otherwise, depending on the state of the soul. This happens over and over again, a continuous cycle of rebirth after death and is dependent on one's karma, basically whether one has been good or bad in the previous life. The process of reincarnation for Hindus is a form of second chances to do better, redemption, with the ultimate goal to reach a state of Nirvana.

A researcher by the name of Dr. Ian Stevenson, one-time Professor of Psychiatry resident at the University of Virginia School of Medicine and once having held the chair in the Department of Psychiatry and Neurology, researched deep into the idea of evidence of reincarnation until his demise. In having done this, he stated that he had found over 3,000 cases of reincarnation.

In one of his papers, *Birthmarks and Birth Defects Corresponding to Wounds on Deceased Persons'*, he used facial markers to ascertain similarities for those who claimed having been incarnated and this also included indicative birthmarks which showed in places where the person claimed to have sustained injuries or even a death wound in a prior life.

His research stated that there were:

"about 35 per cent of children who claim to remember previous lives [who] have birthmarks and/or birth defects that they (or adult informants) attribute to wounds on a person whose life the child remembers."

Others have found similar results. In quite a number of cases, some very young children have such memories of a prior life. Some as young as 18 months, others as old as 5 years talk about past life memories. Along with this, they often exhibit idiosyncrasies or behaviors, including predispositions for certain things, as well as in some cases even inexplicable phobias, which at such a young age and in the circumstances which they live, seemed markedly strange.

Yet, the memories often do gibe and quite accurately to events in the lives of those prior incarnations the children claim to have had. Again, some children even bear birthmarks or perhaps birth defects in shapes and locations that the person in the prior life had wounds, even bullet wounds. In many instances such cases have been confirmed by postmortem evidence of the prior person's death. As children age, some may have these memories but for the most part, they fade by seven years of age. However, such cases have been researched around the world and with some astonishing results.

Here are some such cases:

Ma Tin Aung Myo of Burma: In one particular instance, a case Stevenson became aware of, was about a Burmese girl named Ma Tin Aung Myo. She insisted she had been a Japanese soldier killed in the Second World War. This case is remarkable because it includes large cultural differences between the claimant and the person she once claimed to have been. This is not commonly reported.

Burma was invaded and occupied by the Japanese early in the war. Counterstrikes by allies included heavy bombing. The village of Na-Thul was repeatedly attacked in this way and caused much damage and distress for the villages and their occupants. On top of this, the villages had little choice but to kowtow to their conquerors or die resisting.

Daw Aye Tin the future mother of Ma Tin Aung Myo often spoke with a Japanese army cook who was stationed in the village. Long after the war ended, in 1953, Daw became pregnant. At the time, she kept having a repeating dream concerning the Japanese cook. She hadn't heard from him in years. In the dream he kept saying he would be coming to stay with her family. Later in 1953, Daw had a daughter whom she named Ma Tin Aung Myo. The baby had a small birthmark near her groin.

As the girl aged her parents noticed she had a very real fear of planes. U Aye Maung, her father, was curious about this and wondered why because the war had ended long before and well before the girl had been born. His daughter suffered an ongoing depression and said she wanted to "go home." Later, her father found out that by "home" she meant to be able to go to Japan. He asked her why, and his daughter said it was because she remembered being a Japanese soldier stationed in their village of Na-Thul. She said she had been killed by gunfire from a plane, ostensibly the reason she was afraid of aircraft.

Ma Tin Aung Myo grew, and as she did she remembered other things. She told Ian Stevenson her memories of her previous incarnation as a man who lived in North Japan, that he possessed five children, the oldest being a boy.

She also remembered she had been an army cook. From that point forward, past life memories became more involved and on target. She told them that when she was that soldier, she had been spotted by an enemy plane while wearing shorts and standing by a tree. The plane began firing and so, as the soldier, she ran, but was struck by a bullet in the groin, which killed him. She said the plane had two tails. This was later determined to be a Lockheed P-38 Lightning, a plane used in Burma by the allies.

In the early '70's she had three interviews regarding her reincarnation memories with Ian Stevenson. She said that she wanted to be married to a woman and that she did not like the Burmese weather or spicy food. All the possible characteristics of a Japanese soldier's likes and dislikes of local conditions while serving in a foreign country, such as Burma.

Paddy Fields: Another of Stevenson's cases was about a Sri Lankan girl who said she remembered a life that ended when she had drowned in a paddy field. She described that a bus had driven past and splashed her with water just before she died. Later, it was discovered that a girl in a village not far away had drowned after she had moved to avoid being hit while traversing a narrow highway along the edge of paddy fields. She had fallen into the flooded paddy into deeper water and had drowned. Always, even as a Sri Lankan young girl, she had been oddly fearful of buses and became deeply agitated if near water of any depth at all.

Swarnlata Mishra: Swarnlata Mishra, was born in 1948, and was three when she began having memories from another life, those of a Biya Pathak, a girl who lived about a hundred miles from Swarnlata's village of Madhya Pradesh. She described Biya's home as having been painted white and consisting of four rooms.

She sang songs she used to "know" from that other life and even executed intricate dances her current family and associates had never seen or taught her. Her father recorded some of what she said and began researching evidence of her having been reincarnated.

One researcher discovered that a woman matching Swarnlata's accounts of Biya had passed just 9 years before. Further research showed that a girl called Biya had lived in a house in that village and it matched Swarnlata's descriptions remarkably well.

Hearing this, her father took Swarnlata to the village and had her meet members of the dead Biya's family. As a failsafe to prove she was really remembering a past life she also, unknowingly, met people who weren't family of Biya's but pretended to be. Swarnlata knew they were not part of Biya's family. She wasn't tricked by the ploy. However, she did remember real relatives of the dead Biya's family.

Patrick Christenson/Brother: Patrick Christenson, born in March of 1991 in Michigan, had an older brother, Kevin, who had died 12 years before of cancer. Patrick was born with a diagonal birthmark with a small cut on the right side of his neck. This was the same spot as Kevin's chemotherapy scar. Patrick also had a nodule on his scalp just above his right ear and a clouding of his left eye, even as Kevin had been blinded by cancer in the same eye. When he started walking Patrick had a pronounced limp, even as Kevin had.

At just over four years of age, Patrick told his mother he wanted to return to their old orange and brown house. This was the house the family had lived at when Kevin was still alive. Later, he asked his mother about his surgery and if she remembered it. She replied in the negative because Patrick had never had any surgery. He then indicated the spot above his right ear. There is more about Patrick and the claims he made about Kevin that were accurate, but this suffices here for our purposes as yet another example.

Sam Taylor: Sam Taylor, an eighteen-month-old boy, as his diaper was being changed by his father one day, said, "When I was your age I used to change your diapers." Later the still-virtually an infant child divulged more facts about his grandfather. These turned out to be spot on.

At one point, he even said "his sister" had become a fish. His father knew that the boy's grandfather's sister had been murdered and dumped in the ocean. A telling coincidence, and the boy also knew his grandfather had liked milkshakes made by his grandmother. Sam's grandfather had died well before he had been born and there was no way any of this had been discussed in front of the boy at any point in his short young life, so there was no way the child could have known these things.

When four years of age, Sam saw old pictures spread before him on a table. He easily picked out his grandfather and said: "That's me!" He was correct. Attempting to test him further, his mother showed him a class photograph of his grandfather as a boy. Sixteen other boys were also in the picture. Sam pointed to one of them, saying that it was him (as his grandfather) and this turned out to be correct, as well.

Chapter Conclusion. There are more examples of children having memories of a past life, many more. In fact, there appear to be hundreds to even thousands and the number keeps growing. Many adults, too, claim to have memories, odd phobias since childhood, or attractions to certain places and/or people that can't explain, but which the feel the "sort of remember," or somehow are drawn to, as well. Have you ever had such a feeling, a flash of memory, of felt an odd draw to something or someone since you were a child, and it's a feeling or sensation for which you can't logically or easily account?

This is not an uncommon phenomenon, and scientists simply can't explain it. I know I've had such "moments" of memories, as well, and although I can't pronounce them as having been from a prior life, neither can I logically account for them in this one. Some are just flashes of places or events that I know I couldn't have seen in this life, being in someplace, time, or condition that simply wasn't anywhere I ever could have been. Is this proof of reincarnation for me? I'm not sure, but I find such flashes of memory more than intriguing.

Perhaps, you as a reader of this book, might have also experiences such odd moments of memories, or strong feelings you have been someplace before, when you know you couldn't have been in your current life? This phenomenon, again, is far more common than people realize and instead of being dismissed by those experiencing it as just some sort of fluke or momentary aberration, perhaps such people should report them to researchers in the field of reincarnation and delve deeper into them. Any knowledge gained thus, would be to everyone's benefit in find the answers as to whether reincarnation is real, and if so, thus proving the immortality of the human soul or consciousness.

CONCLUSION

We have explored various forms of how immortality might be possible based on the ongoing research of people in various scientific fields, as well as more exotic ones, as well. One of the strangest yet most promising ideas seems to be quantum immortality. If our universe is a multiverse with many parallel worlds branching off from it, then it is quite likely that all versions of us may exist "out there" somewhere, perhaps even to the point of infinity and maybe even for eternity, as well.

Some of those other versions of "you" would be leading the exact same lives as you are, and others might be leading slightly or majorly different lives that diverge from your own. Quantum immortality is just one way this might be possible. As we've seen, there are a number of other theories on the universe, mainstream ones that also allow for this idea as we have shown in this book. If any of them prove to be true, then so does immortality of varying versions of each of us appear to be true.

Everything from ghosts and other paranormal experiences would then also be explained by such a theory. The human consciousness, if capable of surviving the body's death, could account for the apparitions that people have reported throughout time, and this would also explain many other types of paranormal manifestations, perhaps, as well.

Furthermore, although there is no smoking gun for reincarnation, there does seem to be a great deal of evidence that certainly supports the idea. Have we lived past lives? Have you ever had a memory you simply can't account for? Again, many people have. With the available cases of reincarnation numbering in their hundreds and thousands, perhaps one shouldn't be so quick to ignore or dismiss odd memories one might have as being nothing more than a mere oddity?

We've also explored the idea of the hypersoul, one great soul that transcends all the possible versions of us in the multiverse. If the theory of the multiverse is true, as many scientists think than the you that you know is just a fraction of a much greater you, as one side of a large sparkling diamond is just one facet of a much bigger jewel. That greater soul may go on to be part of an even bigger community of souls that could make up that thing many people call the "cosmic oneness."

The upshot of all this research and investigation seems to be that there is some very real and tangible evidence that one might survive death, that there is something "more" to come and this life isn't "all of it." The consciousness could well go on, and even be immortal, via one of the theories included in this book, or others. And this idea of an immortal consciousness isn't a new idea. After all, various religions and mystics around the word have been saying this idea of an immortal consciousness and/or a cosmic consciousness for thousands of years now. Just maybe, they were right all along.

There does seem to be a strong trend in science, particularly quantum physics and cosmology as well, toward answers that sound almost metaphysical in nature. Probability waves that allow for realties where anything that is possible has to happen? A multiverse composed of countless universes, all with their own possibilities, ones which may actually exist? The idea or theory of quantum immortality actually being real? All these lead us to wonder if science and religion are not having, after all these centuries, some sort of rapprochement, that two sides of an ancient wheel might actually be coming together at last? As the famous poet, Edgar Allan Poe once wrote:

I have reached these lands but newly
From an ultimate dim Thule—
From a wild weird clime that lieth, sublime,
Out of SPACE—Out of TIME.

From Edgar Allan Poe's poem, Dreamland

END

REFERENCES

https://www.sciencealert.com/science-discovers-human-brain-works-up-to-11-dimensions

https://www.scientificamerican.com/article/multiverse-the-case-for-parallel-universe/

https://www.encyclopedia.com/philosophy-and-religion/other-religious-beliefs-and-general-terms/religion-general/automatic-writing

https://www.google.com/search?client=firefox-b-1-d&q=is+a+jiffy+the+smallest+unit+of+time%3F

https://www.scientificamerican.com/article/does-consciousness-pervade-the-universe/

https://bdtechtalks.com/2018/12/03/jeremy-howard-ai-deep-learning-myths/human-brain-digital-x-ray-3d-rendering-2/

https://blogs.scientificamerican.com/sa-visual/the-beautiful-complexity-of-the-cosmic-web/

https://www.ncbi.nlm.nih.gov/pmc/articles/PMC6172100/

https://www.catholicherald.com/Faith/Near-death_experiences__evidence_of_afterlife/

https://www.ncbi.nlm.nih.gov/pmc/articles/PMC6172100/

https://www.academia.edu/39577607/The_Consciousness_Revolution_in_Science?email_work_card=view-paper

https://www.academia.edu/39577607/The_Consciousness_Revolution_in_Science?email_work_card=view-paper

https://www.scientificamerican.com/article/what-is-consciousness/

https://www.nature.com/articles/d41586-019-02207-1

https://www.theguardian.com/science/2015/jan/21/-sp-why-cant-worlds-greatest-minds-solve-mystery-consciousness

https://en.wikipedia.org/wiki/Wikipedia:Contents/Religion_and_belief_systems

https://en.wikipedia.org/wiki/Wikipedia:Contents/Religion_and_belief_systems

https://medium.com/intercultural-mindset/belief-systems-what-they-are-and-how-they-affect-you-1cd87aa775ff

https://www.historyhaven.com/BELIEF%20SYSTEMS.htm

https://en.wikipedia.org/wiki/The_Quantum_Universe

https://ucsdnews.ucsd.edu/pressrelease/scientists-explore-signals-for-a-quantum-universe

https://www.nature.com/articles/434438b

https://www.popularmechanics.com/science/a36329671/is-the-universe-conscious/

https://www.npr.org/sections/13.7/2017/07/12/536752502/is-the-universe-conscious

https://owlcation.com/stem/Is-the-Universe-Conscious

https://owlcation.com/stem/Is-the-Universe-Conscious

https://www.space.com/32728-parallel-universes.html

https://www.treehugger.com/parallel-worlds-exist-and-interact-with-our-world-say-4863488

https://www.newscientist.com/article/mg24532770-400-we-may-have-spotted-a-parallel-universe-going-backwards-in-time/

https://www.npr.org/sections/13.7/2011/08/23/139875744/defining-the-universe-harder-than-you-think

https://www.nationalgeographic.com/science/article/origins-of-the-universe

https://www.quantamagazine.org/physicists-debate-hawkings-idea-that-the-universe-had-no-beginning-20190606/

https://www.bbc.com/future/article/20200117-what-if-the-universe-has-no-end

https://en.wikipedia.org/wiki/String_theory

https://www.space.com/17594-string-theory.html

https://www.livescience.com/65033-what-is-string-theory.html

https://en.wikipedia.org/wiki/Holographic_principle

https://www.scientificamerican.com/article/information-in-the-holographic-univ/

https://www.quantamagazine.org/how-our-universe-could-emerge-as-a-hologram-20190221/

https://en.wikipedia.org/wiki/Quantum_suicide_and_immortality

https://en.wikipedia.org/wiki/Quantum_suicide_and_immortality

https://knappily.com/ethics/quantum-suicide-you-will-live-forever-261

https://www.forbes.com/sites/startswithabang/2017/10/14/ask-ethan-is-the-universe-finite-or-infinite/

https://phys.org/news/2015-03-universe-finite-infinite.html

https://www.discovermagazine.com/the-sciences/where-is-the-edge-of-the-universe

https://www.esa.int/Science_Exploration/Space_Science/Is_the_Universe_finite_or_infinite_An_interview_with_Joseph_Silk

https://www.britannica.com/topic/immortality

https://aeon.co/essays/theres-a-big-problem-with-immortality-it-goes-on-and-on

https://library.wcupa.edu/c.php?g=61498&p=395636

https://credoreference.libguides.com/c.php?g=139767&p=915791

https://en.wikipedia.org/wiki/Big_Bang

https://en.wikipedia.org/wiki/Quantum_entanglement

https://www.sciencedaily.com/terms/quantum_entanglement.htm

https://www.google.com/search?q=quantum+entanglement&client=firefox-b-1-d&sxsrf=ALeKk02WXOsxDdlN_7hC9ARXLZAk_BC7rA%3A1629564813678&ei=jS8hYefVKMiw5NoPlaWNqAw&oq=quantum+ent&gs_lcp=Cgdnd3Mtd2l6EAEYADIECCMQJzIKCAAQgAQQhwIQFDIFCAAQgAQyBQgAEIAEMgUIABCABADIFCAAQgAQyBQgAEIAEMgUIABCABADIFCAAQgAQyBQgAEIAEOgcIABBHELADOgUIABCRAjoFCC4QkQI6EQguEIAEELEDEIMBEMcBENEDOg4ILhCABBCxAxDHARDRAzoICC4QgAQQsQM6DgguEIAEELEDEMcBEKMCOgIABBDOgcIABCxAxBDOgsILhCABBDHARDRAzoICAAQgAQQsQM6BQguEIAESgQIQRgAULIKWLAaYMEzaABwA3gAgAGCA4gBgxOSAQcxLjYuMy4ymAEAoAEByAEIwAEB&sclient=gws-wiz

www.ingramcontent.com/pod-product-compliance
Lightning Source LLC
Chambersburg PA
CBHW052330220526
45472CB00001B/347